# CHEMISTRY OF LOWER HETEROCYCLIC COMPOUNDS

# CHEMISTRY OF LOWER HETEROCYCLIC COMPOUNDS


Dr. Mahima Srivastava, M.Sc. [Gold Medalist], Ph.D.
Associate Professor,
Department of Chemistry,
D.B.S. (PG) College, H.N.B. Garhwal Central University,
Uttarakhand, India


*CWP*

**Central West Publishing**

**Disclaimer**
Every effort has been made by the publisher, editors and authors while preparing this book, however, no warranties are made regarding the accuracy and completeness of the content. The publisher, editors and authors disclaim without any limitation all warranties as well as any implied warranties about sales, along with fitness of the content for a particular purpose. Citation of any website and other information sources does not mean any endorsement from the publisher, editors and authors. For ascertaining the suitability of the contents contained herein for a particular lab or commercial use, consultation with the subject expert is needed. In addition, while using the information and methods contained herein, the practitioners and researchers need to be mindful for their own safety, along with the safety of others, including the professional parties and premises for whom they have professional responsibility. To the fullest extent of law, the publisher, editors and authors are not liable in all circumstances (special, incidental, and consequential) for any injury and/or damage to persons and property, along with any potential loss of profit and other commercial damages due to the use of any methods, products, guidelines, procedures contained in the material herein.

A catalogue record for this book is available from the National Library of Australia

ISBN (print): 978-1-922617-36-1

I dedicate this book to my parents, brother Anuj Anand,
son Daksh Srivastava and my teachers

J N C A S R

प्रोफेसर सी. एम. आर. राव, F.R.S.

Professor C.N.R. Rao, F.R.S.
Linus Pauling Research
Professor and
Honorary President

जवाहरलाल नेहरू  Jawaharlal Nehru
उन्नत वैज्ञानिक  Centre For Advanced
अनुसंधान केंद्र  Scientific Research

## FOREWORD

Majority of natural products comprise of heterocyclic rings contributing to extensive structural diversity. Along with this they often possess remarkable biological activity. Medicinal chemists have espoused this property in designing most of the small-molecule drugs in use today. This book offers readers a fundamental understanding of the details of lower heterocyclic compounds and their applications in diverse areas in addition to natural products such as amino acids, DNA, vitamins, and antibiotics.

Thus, the importance of heterocyclic chemistry continues to increase and the book by Dr. Mahima Srivastava is a welcome addition to the available literature on the subject. Its scope places it in a useful niche among all the learners of this category. The author has retained the tested and tried classical approach but has successfully placed her own individual spin on her arrangement. She has put together a selected range amongst the most important of the vast array of content available. The matter is ordered in a clear, logical, and lucid pattern over the whole book. The book includes content dealing with comprehensive aspects of lower heterocycles such as Aziridine, Oxirane, Thiirane, Azetidine, Oxetane, Thietane and five – membered heterocyclic compounds and their benzo – fused derivatives. Applied aspects concerning these compounds have also been taken care of in detail.

The present work should be of value to the students and practitioners of heterocyclic chemistry at all levels. It is also recommended to researchers working in the fields of heterocyclic chemistry and students in the field of organic chemical science. It will be of assistance in presenting a crisp and modern view of the subject to those who use heterocyclic chemistry.

I heartily congratulate Dr. Mahima Srivastava for the magnificent job she has done.

C N R Rao

# About the Author

Dr. Mahima Srivastava is currently working as Associate Professor in Department of Chemistry of Dayanand Brajendra Swarup Post Graduate College, Dehradun. A Ph.D. in chemistry, she is a gold medalist ranking first in the order of merit in Master of Science from CSJM University, Kanpur. She is an astute and result oriented professional and has a long experience of teaching undergraduate and postgraduate students. She has been conducting research activities alongside to keep abreast with the current. In addition to this, she holds various responsibilities of the organization viz., conducting seminars/symposia/workshop/lecture series, paper setter at post graduate level of various universities, evaluation of university answer scripts and conducting university examination.

# Preface

I feel immense pleasure in presenting this book to the teachers and students of undergraduate and postgraduate level. This book is the outcome of my long teaching experience of more than a decade at graduate and postgraduate level.

A long span of interaction with students has guided me to give special attention to dark areas which students find difficult to understand. The book has been written in a simple and easy-to-understand language. The topics pertaining to the syllabus have been covered to full extent. Some salient features of the book are:

- Many illustrations and examples have been provided to make the reading interesting and engaging.
- Maximum methods of synthesis and reactions for every compound have been discussed.
- The Compounds in every category have been discussed in detail for clear understanding.
- General methods of synthesis and reactions has also been elaborated along with specific methods of synthesis and reactions so that basic pattern in which the compound behaves may be followed.
- The uses of every compound have been mentioned and general applications have been discussed in a separate chapter.

I am thankful to the management and editorial team of Central West Publishing for the help and support in the publication of this book.

I believe that the book contains all that is needed to understand the subject in a systematic manner. However, I also believe that no work is perfect and there is always scope for improvement. I would welcome suggestions from the teaching fraternity and students for further improvement of the book.

# Symbols and Abbreviations

Å = Angstrom is a metric unit of length equal to 10−10 m. The unit is named after the Swedish physicist Anders Jonas Ångström

D = Dipole Moment

$\alpha$ and $\beta$ carbons = The first carbon atom after the carbon atom bonded to a functional group is known as alpha carbon and further carbon in a chain are known as beta, gamma, etc.

$\lambda_{max}$ = wavelength of maximum absorbance

$\varepsilon$ = molar absorptivity or molar absorption coefficient

Me = Methyl functional group

Et = Ethyl functional group

Ph = Phenyl functional group

IUPAC = International Union of Pure and Applied Chemistry

I = Inductive effect

M = Mesomeric effect

# Table of Contents

# CHAPTER 1

## INTRODUCTION

Heterocyclic compounds find a widespread usage as their structures can be manipulated to achieve a required modification in function. They are those cyclic compounds in which one or more of the ring carbons are replaced by another atom. The non-carbon atom in such rings is referred to as heteroatom. These compounds have a varied range of applications:

- Pharmaceuticals
- Corrosion Inhibitors
- Dyes and pigments
- Intermediates in organic synthesis
- Antioxidants

An important feature of the structure of many heterocyclic compounds is that it is possible to incorporate functional groups either as substituents or part of the ring system itself. This means that the structures are particularly versatile as a means of providing a functional group.

An interesting feature of these compounds is that they are finding an increasing use as intermediated in organic synthesis as a relatively stable ring system can be brought about through several steps and then interfered at the required stage for other functional groups as shown below.

Many asymmetric synthetic reactions accomplish complexes derived from heterocyclic compounds being used as chiral ligands for transition metals as catalysts. The nucleic acid bases, **Pyrimidine and Purine** ring systems, which form the basis of replication in biological systems are also heterocyclic compounds. Other natural systems comprising of heteroatoms in carbon ring systems are chlorophyll, heme, vitamin B-complex and amino acids.

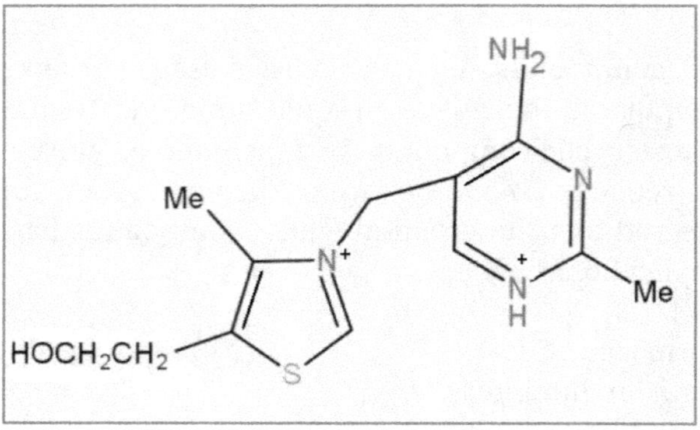

**Vitamin B1 (Thiamine) Structure**

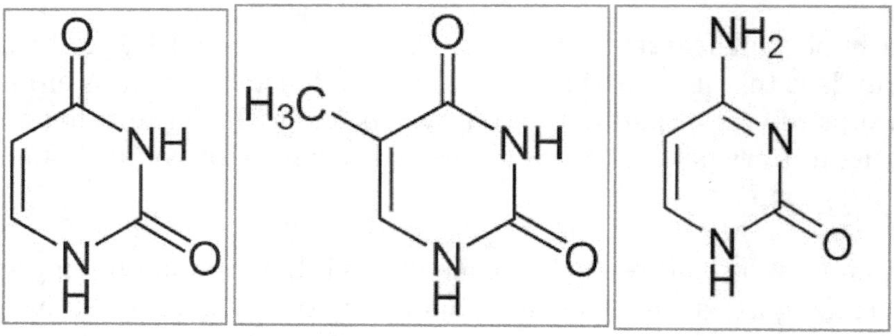

**Structure of Pyrimidine Bases (Uracil, Thymine and Cytosine)**

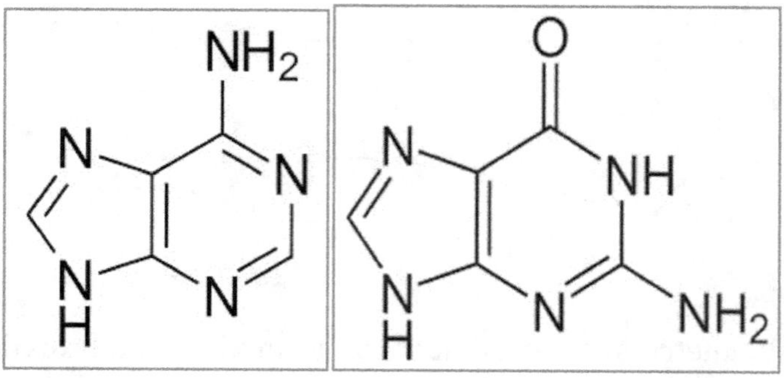

**Structure Purine Bases (Adenine and Guanine)**

In the same way, heterocyclic compounds are used largely as drugs and medicines. For example, **Ergotamine** is used for treating migraine; **Ranitidine** treats peptic ulcers; **Zidovudine** is used in the cure of AIDS; hairy cell leukemia is treated with **Pentostatin**; etc.

**Zidovudine**

**Ranitidine**

## REFERENCES

Czarnik, A. W. (1996) Acc. Chem. Res, 29, 112.
Comprehensive Heterocyclic Chemistry (1984) Vol. 1, ed. O Meth–Cohn, Pergamon Press, Oxford.
Clough, J. M. and Godfrey, C. R. A. (1995) Chem. Br., 466.
Jones, M. F. (1988) Chem. Br., 1122.
Elion, G. B. (1989) Angew. Chem, Int., Ed. Engl., 28, 870.
Durant, G. J. (1985) Chem. Soc. Rev, 14, 375.

# CHAPTER 2

## PROPERTIES OF HETEROCYCLIC COMPOUNDS

Aromatic systems have been classified into two types:

- **Carbocyclic** – The cyclic organic compounds containing all carbon atoms in ring formation are carbocyclic aromatic systems.
- **Heterocyclic**– The cyclic organic compounds containing at least one atom other than carbon atom in ring formation are heterocyclic aromatic systems.

## 2.1 STRUCTURAL PROPERTIES

### 2.1.1 STRUCTURE OF MONOCYCLIC HETEROATOMIC COMPOUNDS

Pyridine is different from benzene by replacement of CH by N. This replacement leads to the loss of regular hexagonal geometry; shorter carbon-nitrogen (C-N) bond and a lone pair of electrons in place of hydrogen.

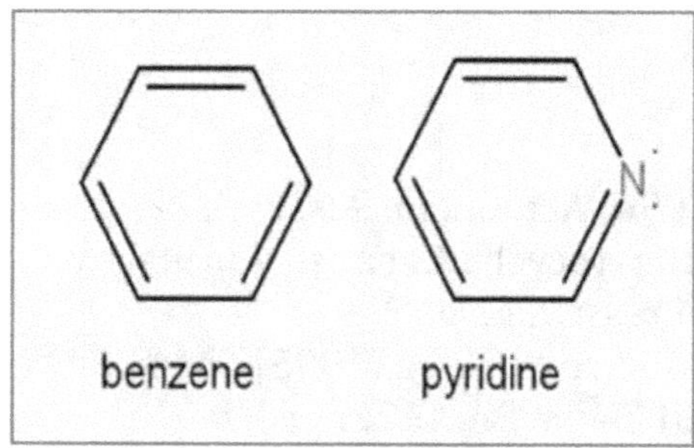


The presence of electronegative nitrogen leads to the development of a strong permanent dipole.

## BOND LENGTH

**C-C** SINGLE BOND – 1.40Å
**C-C** DOUBLE BOND – 1.39Å
**C-N** SINGLE BOND – 1.34Å

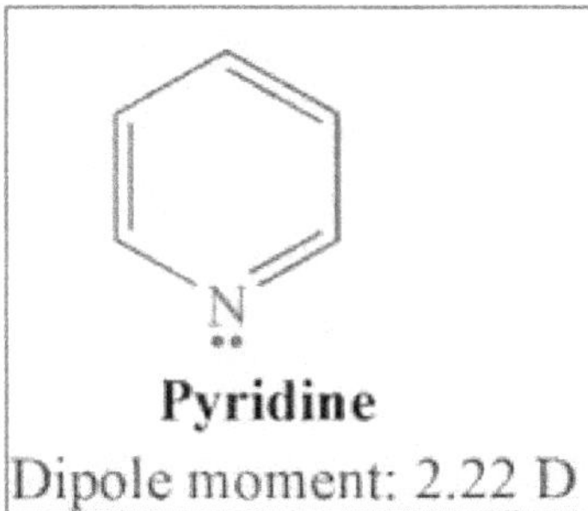

**Pyridine**
Dipole moment: 2.22 D

**Diazines** have a similar structure but with a difference of two nitrogen atoms and hence two lone pairs.

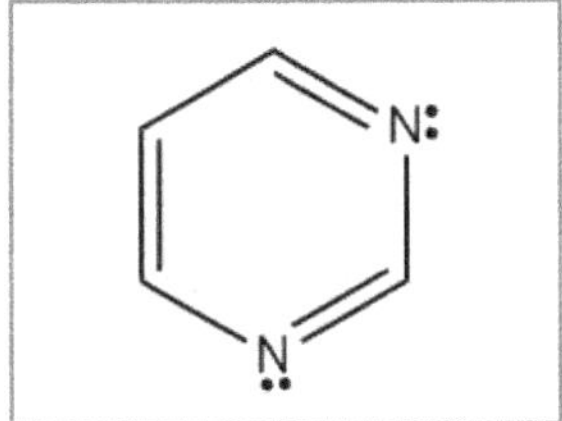

In **pyrylium**, the lone pair of electrons is carried by oxygen. The positive charge gets delocalized on the shown carbon atoms.

**Pyridine** with some oxygen at either position 2- or 4- is referred to as pyridine; however, pyrones have oxygen as heteroatom in ring structure also.

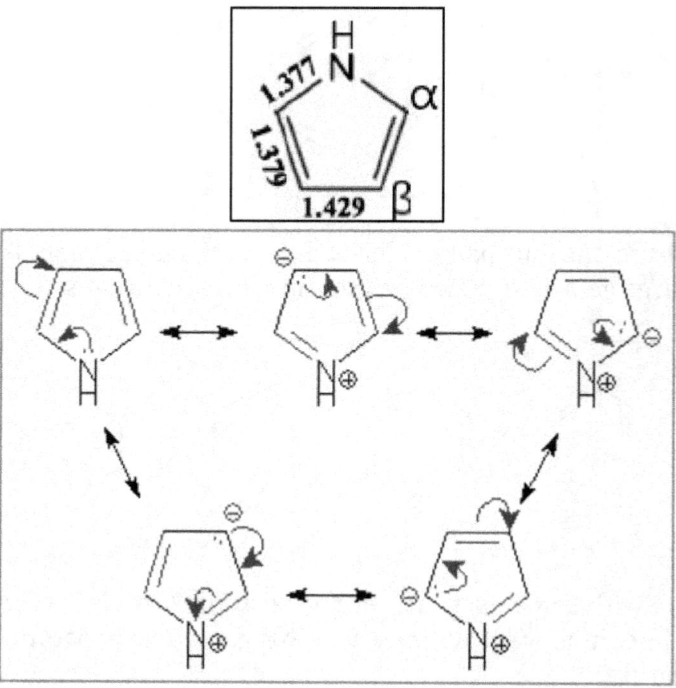

Planar, unsaturated heterocycles containing five atoms should also be discussed here. Pyrrole has N-atom sp$^2$ hybridized with three σ-bonds, two of which are with cyclic carbon–atoms and one with hydrogen. The carbon atom next to nitrogen atom is treated as α– while those far away are β-.

The effect of **resonance** gives rise to a partial positive charge on nitrogen and a partial negative charge on carbons. Furan is a close analogue of Pyrrole in which -NH- is replaced by -O-.

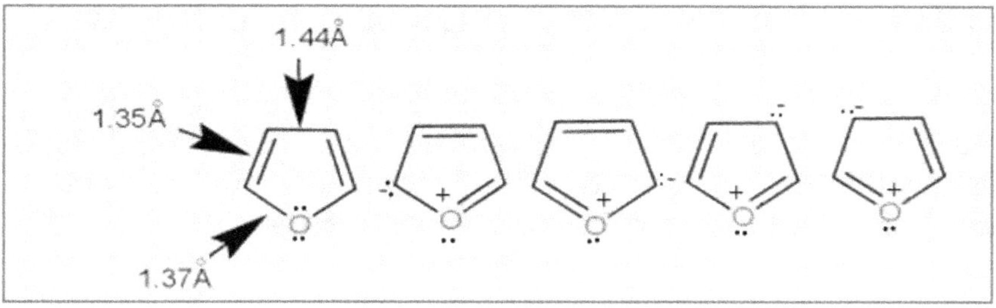

**Thiophene** also belongs to the same series with replacement of -O- by -S-.

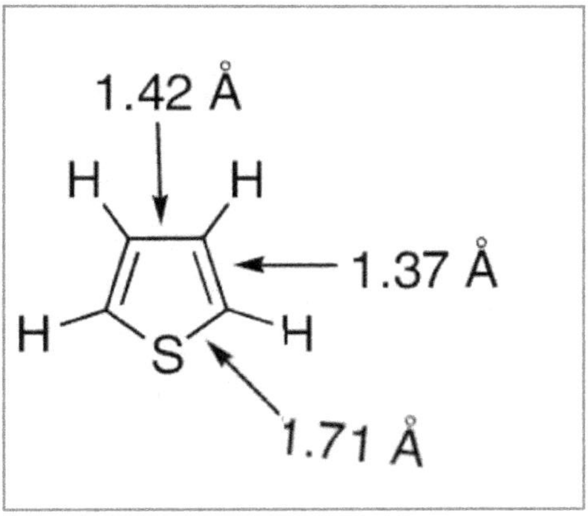

There are many other five membered heterocycles with two or more heteroatoms.

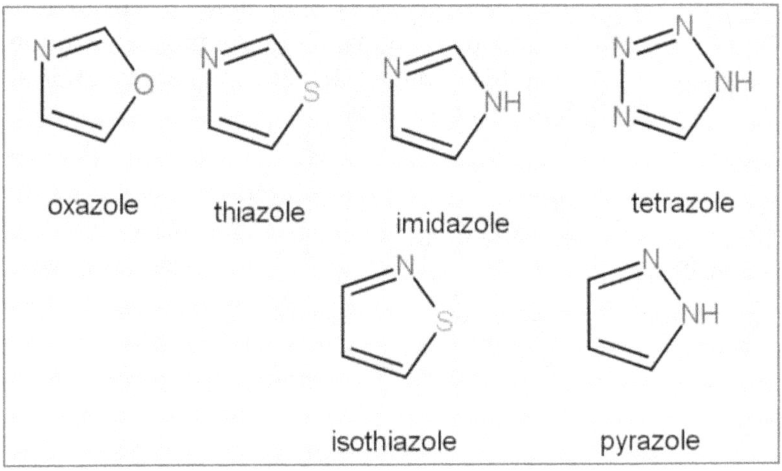

## 2.1.2 STRUCTURE OF BICYCLIC HETEROATOMIC COMPOUNDS

The benzo – fused ring systems have modified aromatic properties. There are two such systems: one where a six – membered ring is fused with a five – membered ring and the other where two six membered rings are fused (the dibenzo – fused systems).

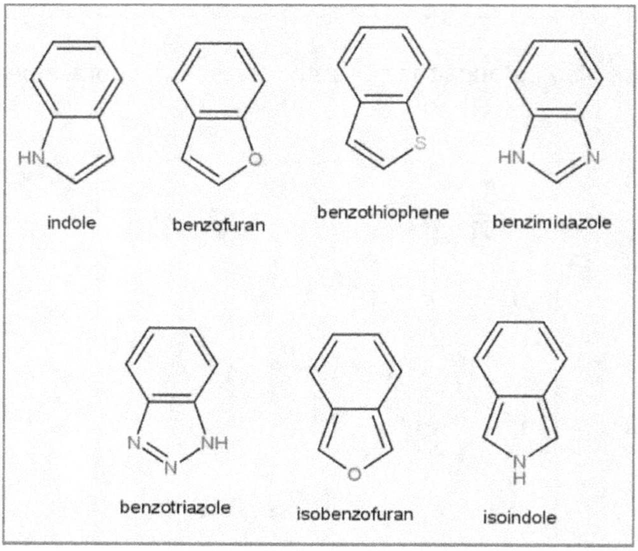

## 2.1.3 MISCELLANEOUS HETEROATOMIC COMPOUNDS

A large number of fused heterocycles can be made by fusion of two monocyclic heteroatomic ring systems. These compounds are classified as aromatic for they have planar structure with total 10 $\pi$-electrons, thus obeying Huckel's rule for n=2.

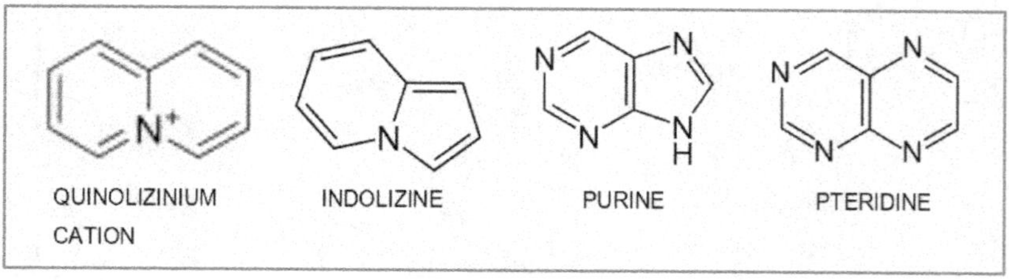

## 2.2 SPECTROSCOPIC PROPERTIES

The spectral data is found to be of great use in the analysis of structures of reactive species and intermediates and in the understanding of new reactions. The **ultraviolet spectroscopic** data of five membered heteroaromatic systems show just one medium to strong low wavelength band, with no fine structure. The bicyclic azines have a more complex electronic absorption, and the n to $\pi^*$ and $\pi$ to $\pi^*$ band overlap. However, the spectra of monocyclic azines shows two bands, each with fine structure: one in the range of 240-260 nm (narrow range) and other at 270-340 nm (broader range).

**Table 1 UV spectra of five – membered heterocycles**

| HETEROCYCLE | $\lambda_{max}$ (nm) | $\varepsilon$ |
|---|---|---|
| Monocyclic | | |
| Pyrrole | 210 | 5100 |
| Furan | 200 | 10000 |
| Thiophene | 235 | 4300 |
| Imidazole | 206 | 3500 |
| Bicyclic | | |
| Indole | 288, 261, 219 | 4900, 6300, 25000 |
| Benzo (b) thiophene | 288, 257, 227 | 2000, 5500, 28000 |
| Benzo (b) furan | 281, 244 | 2600, 11000 |
| Purine | 263 | 7950 |
| Indolizine | 347, 295, 238 | 1950, 3600, 32000 |

The chemical shifts are influenced by the distortion of the $\pi$- election distribution caused by the heteroatoms. Several other factors affect the chemical shift values like effect of solvents.

**Table 2 Proton chemical shift values for aromatic CH**

| Heterocycle (solvent) | $\tau_1$ | $\tau_2$ | $\tau_3$ | $\tau_4$ | $\tau_5$ | $\tau_6$ | $\tau_7$ | $\tau_8$ |
|---|---|---|---|---|---|---|---|---|
| Quinoline (Me$_2$CO) | - | 1.1 | 2.5 | 1.7 | 2.1 | 2.4 | 2.3 | 1.9 |
| Pyrylium (liq.SO$_2$) | - | 0.4 | 1.5 | 0.7 | - | - | - | - |
| Thiophene (C$_6$H$_{12}$) | - | 2.9 | 3.0 | - | - | - | - | - |
| Furan (C$_6$H$_{12}$) | - | 2.7 | 3.8 | - | - | - | - | - |
| Purine (D$_2$O) | - | 1.5 | - | - | - | 1.3 | - | 1.7 |

## 2.3 REACTIVITY OF HETEROATOMIC COMPOUNDS

The reactivity of heterocyclic compounds towards different reagents shows selectivity at specific positions of the ring. There is no generalized explanation for the reactivity of heterocyclic compounds; also, the role of solvent cannot be ignored particularly in the case of five-membered heteroatomic compounds. Heterocyclic compounds are somewhat resistant to oxidation but get possibly reduced. Heterocyclic compounds have evolved greatly in the field of organometallic chemistry. Most important in this series are lithium derivatives.



Normally, **lithiation** is carried out with alkyl lithium; n-Butyllithium being most widely used. Six-membered heterocyclic compounds witness the displacement of a halide by a nucleophile; however, in five membered heterocyclic species such a reaction takes place only under special conditions.

X=electrophile

Electrophilie sustitution at carbon

**Electrophilic reactions** take place at nitrogen as well as at carbon. The lone pair of nitrogen is donated to the electrophile, thereby bringing about addition.

## REFERENCES


Joule, J. A. and Smith, G. F. (1978) Heterocyclic Compounds, Van Nostrand Reinhold Co., 2nd Ed., London.

Beak, P., Covington, J. B., Smith, S. G., White, J. M. and Zeigler, J. M. (1980) J. Org. Chem., 45, 1354.

Fringuelli, F., Marino, G., Taticchi, A. and Grandolini, G. (1974) J. Chem. Soc., Perkin Trans., 2, 332.

Jackman, L. M. and Sternbell, S. (1969) "Applications of NMR spectroscopy in organic chemistry", Pergamon press.

Speranza, M. (1986) Adv. Heterocycl.Chem., 40, 25.

Lyle, R. E. and Anderson, P. S. (1966) Adv. Heterocycl. Chem., 6, 46.

Katritzky, A. R. and Taylor, R. (1990) Adv. Heterocycl. Chem., 47, 1.

# CHAPTER 3

## APPLICATIONS OF HETEROCYCLIC COMPOUNDS

Heterocyclic compounds have numerous applications and have been found immensely useful in almost all the sectors. Although many applications have been discussed alongside the compounds in the chapters, but a clear mention of the left out, yet highly relevant compounds, is a must. The medicine sector, agriculture, dyes and pigments, additives and photography are some of the areas where these compounds render best of their services. So, lets proceed with it.

## 3.1 AS MEDICINES

Heterocyclic compounds have been used in medicines since ancient times. During earlier years, medicines like **Pyrazolone** antipyrine as antipyretic and analgesic (in 1884), **Barbitone** as barbiturate (in 1903) a mixture of Proflavine and 10-methochloride, in the name of **Acriflavine,** as a trypanocide (in 1912), **Methylene Blue** as a weak antiseptic, **Phenazopyridine** as a treatment of infections of urinary tract (in 1926) and compound like **Pamaquine** (Plasmochin), **Mepacrine** (Atebrin) and **Sontochin** were used as cure of malaria.

pyrazolone
(R=H)

barbitone
(R=Et)

methylene blue

proflavine

phenazopyridine

pamaquine

mepacrine

R=CHMe(CH₂)₃NEt₂
R'=Me

sontochin

Various medicines of natural origin are also employed as they contain alkaloids. **Papaverine** (benzyl isoquinoline alkaloid) from *Papaver somniferum* is a smooth muscle relaxant and vasodilator. Thebaine is the main alkaloid from *Papaver bracteatum* is converted into codeine which is used as an antitussive drug. Coughing may also be suppressed by **Oxolamine.** Other similar acting drugs are **Narcotine** (noscapine) and **Phenothiazine Dimethoxanate**.

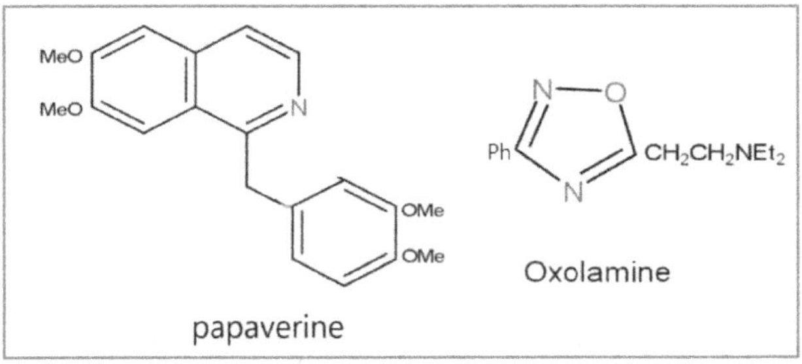

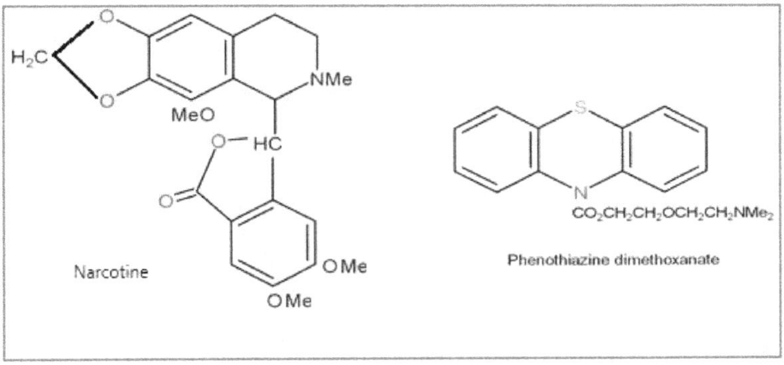

Many alkaloids stimulate muscles and glands like **Pilocalpine,** used to produce pupillary constriction is the eye and diaphoresis (sweating), **Arecoline** is employed as a diaphoretic and Anti-helmintic and **Nicotine** stimulates striated muscles.

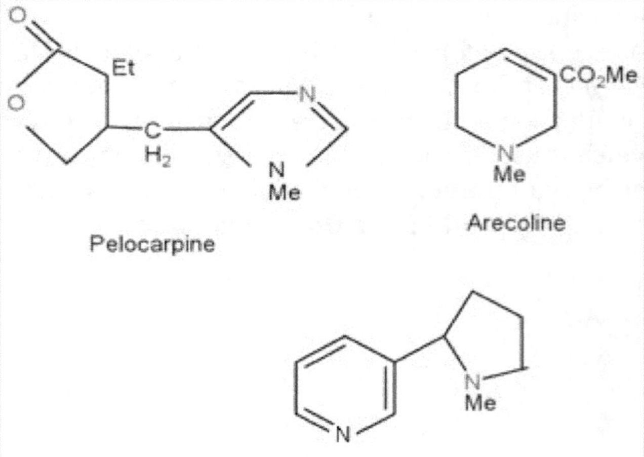

Pelocarpine
Arecoline
Nicotine

Most importantly, xanthine like **caffeine**, **theophylline** and **theobromine** present in tea, coffee and cocoa are stimulants.

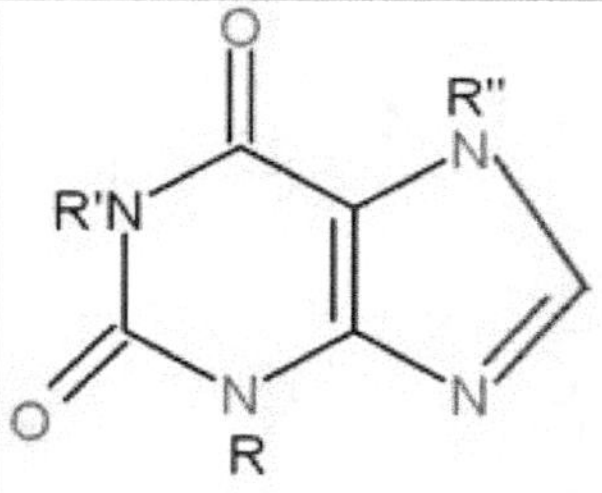

R=R'=R"=Me
caffeine
R=R'=Me , R"=H
theophylline
R=R"=Me, R'=H
theobromine

**Dicoumarol** is used as an anticoagulant in the treatment of thromboses.

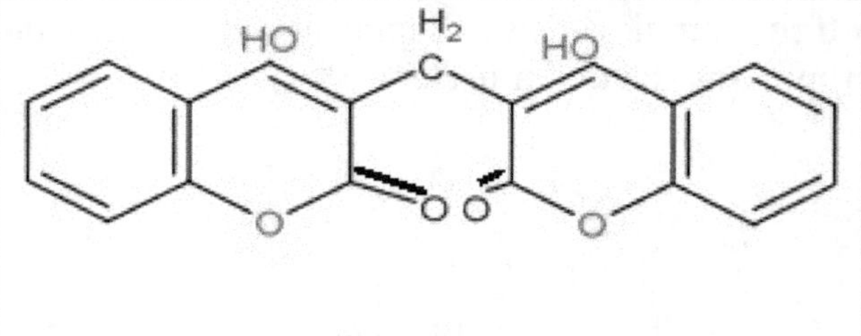

Dicoumarol

Antibiotics are substances that are produced by microorganisms to curb other micro-organism activity or kill them. The earliest antibiotic is 2–pentenylpenicillin to be used as early as 1941. Clavulanic acid is used clinically in mixture with amoxycillin.

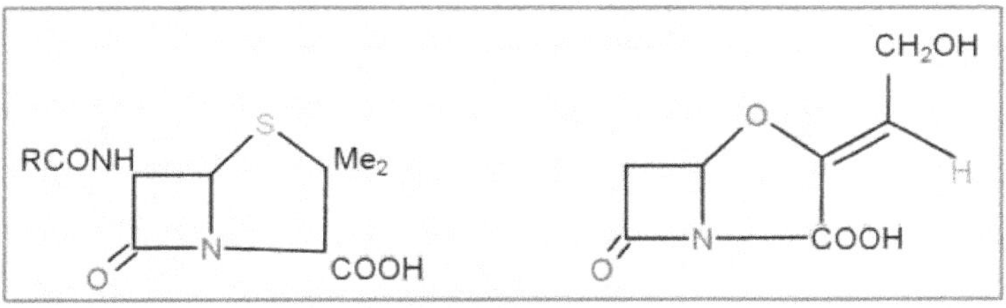

R = 2–pentenyl [2-pentenylPenicillin]
R = α-amino benzyl [Ampicillin]
R = α–amino-p-hydroxy benzyl [Amoxycillin]

Clauvulanic Acid

**Griseofulvin**, an antifungal agent, is useful for treating skin fungal infections.

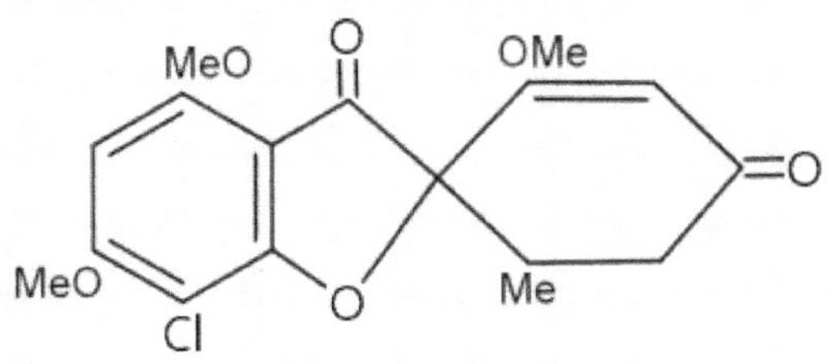

Griseofulvin

Vitamins are one more class that constitutes of heterocyclic compound. They are supplied by various food items or by synthesis by intestinal flora and their deficiency leads to disorders.

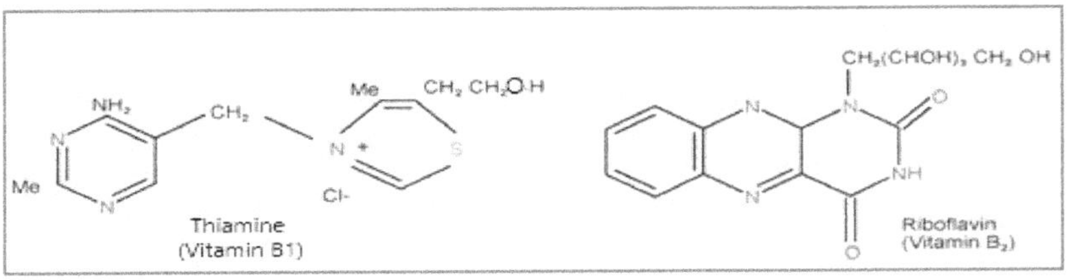

Thiamine
(Vitamin B1)

Riboflavin
(Vitamin $B_2$)

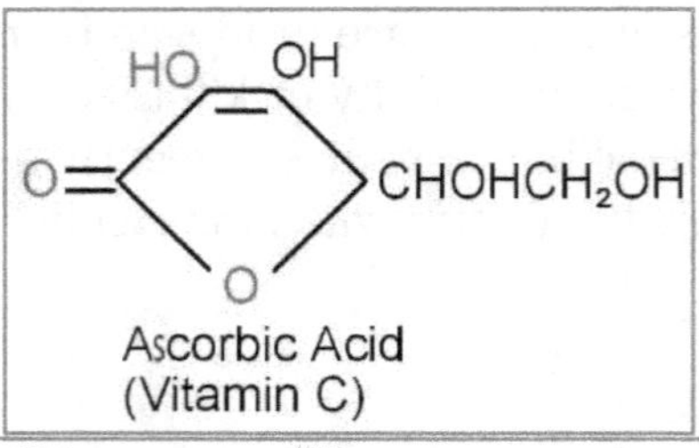

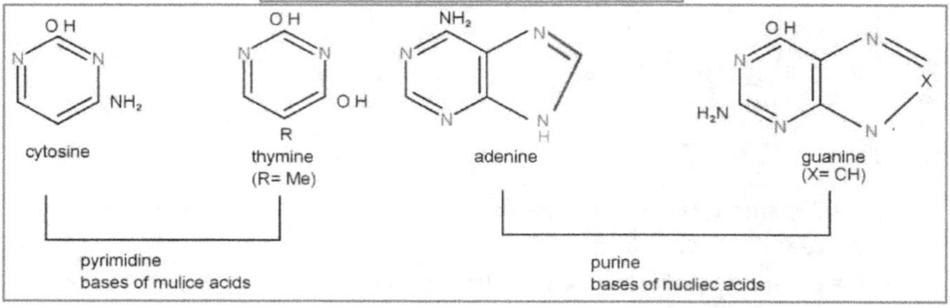

16





Anticonvulsants are drugs which help in curing convulsions for example, **Mephobarbital, Metharbitone, Phenytoin, Succinimide, Ethosuximide, Phensuximide, Methsuximide, Troxidone, Aloxidone, Paramethadione** and **Ethadione**

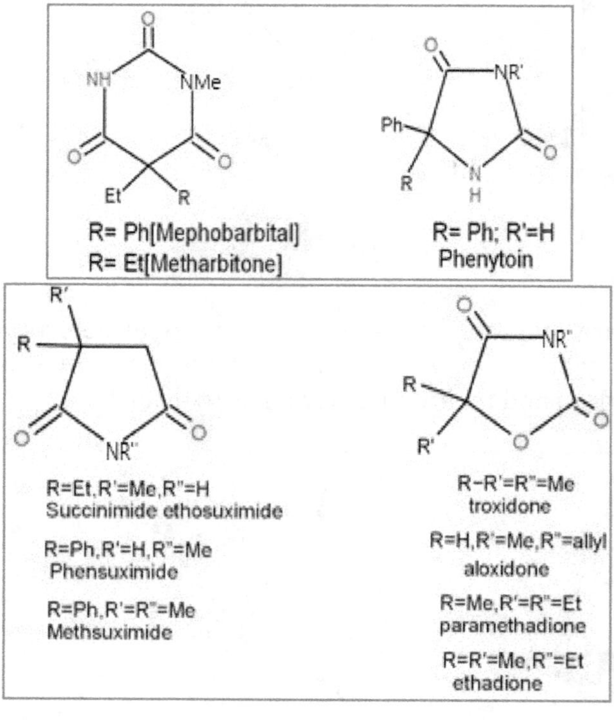

Anti-anxiety drugs like **Diazepam** and **Nitrazepam** have proved to be powerful anticonvulsants.

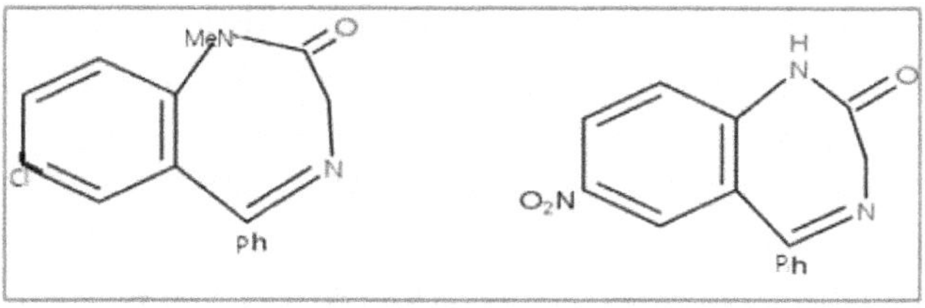

DIAZEPAM                              NITRAZEPAM

Heterocyclic compounds, like **Pethidine,** are a strong analgesic, imidazoline **Clonidine** acts in vasomotor relaxation, **Xanthopterin** affects renal function, **Mepivacaine** as anaesthetic and **Nalidixic Acid** used for urinary tract infections are a few to mention.

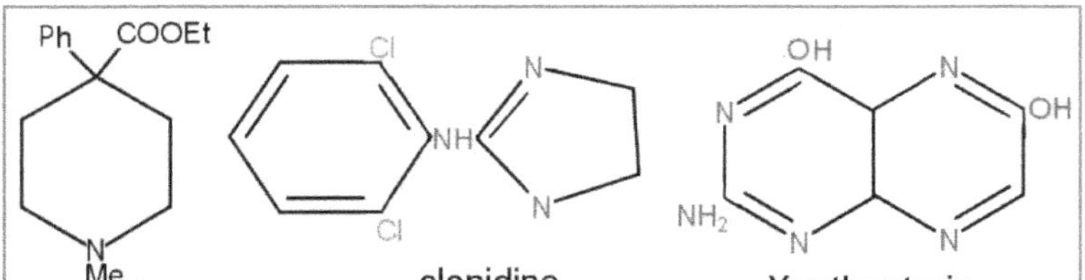

pethidine                clonidine            Xanthopterin

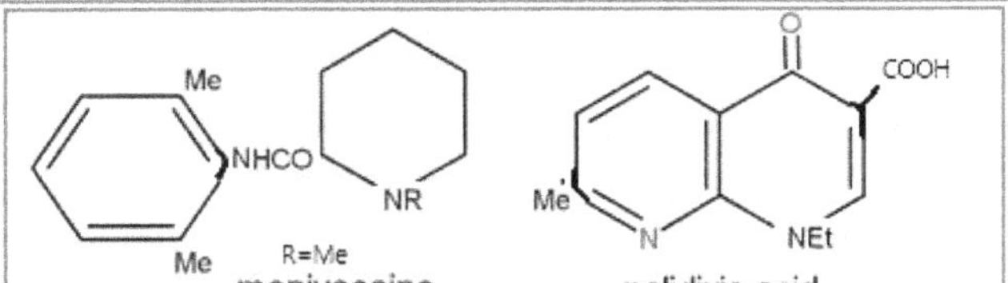

mepivacaine              nalidixic acid

## 3.2 AS CHEMICALS IN AGRICULTURE

Various fungicides, like **Triadimefon** and **Fenarimol** have shown highly effective results.

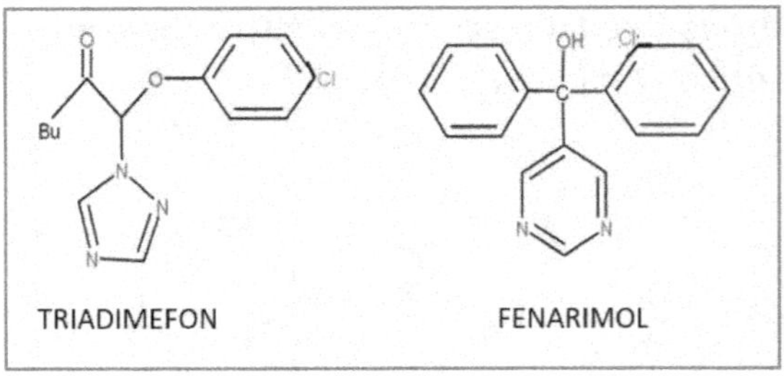

The most widely spread variety is of insecticides like chlorpyrifos, **Diazinon**, **Thionazin** (nematicide), **Quinalphos** (insecticide and acaricide), **Phosalone**, **Methidathion** etc.

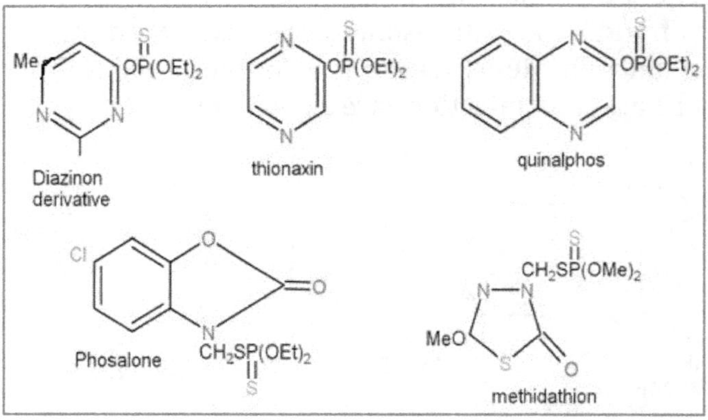

## 3.3 AS POLYMERS

A huge molecule that has a repeating unit containing a heterocycle in the main chain, side chain or terminus is a heterocyclic polymer. These polymers are of different types like polymers formed from heterocyclic monomers, heterocycle-forming polymerization and polymers formed via polymer modifications.

In the first category are the most widespread compounds almost in all kinds of heterocycles. For example, five membered cyclic amide, **Pyrrolidone** (as in vinylpyrrolidone) polymerizes readily (giving polyvinylpyrrolidone). The resulting polymer finds utility as a blood plasma extender in cosmetics and pharmaceuticals.

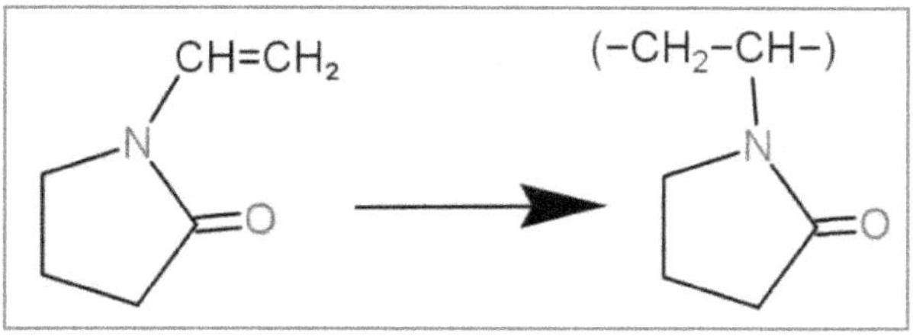


Polymerized **N-vinylpthalimide** is reported to be an excellent photoconductor.

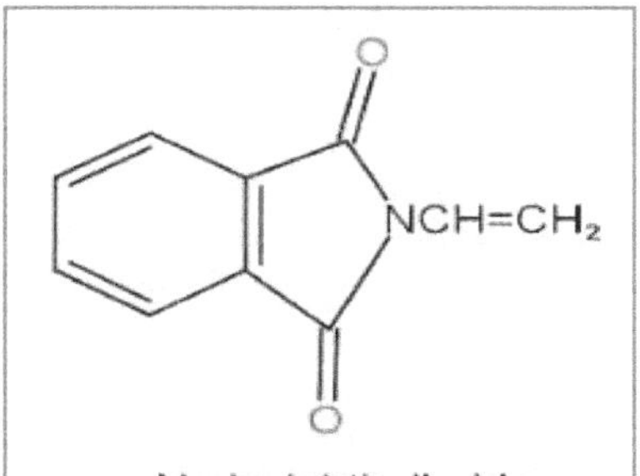


N-vinylphthalimide

**Polyvinyl Indoles** are found to be good insulators and **Polyvinyl Carbazoles** are known for their excellent photoconductivity.

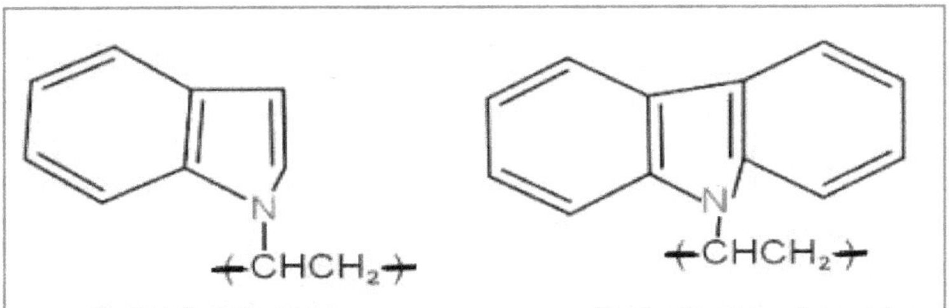


In the second category are **heterocycle-forming polymerization** reaction which have been most extensively reviewed.

20

## a. Pyrroline

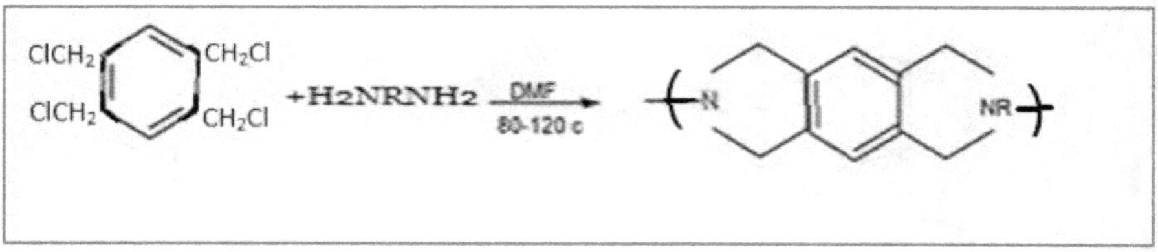

## b. 3-Pyrrolin-2-one

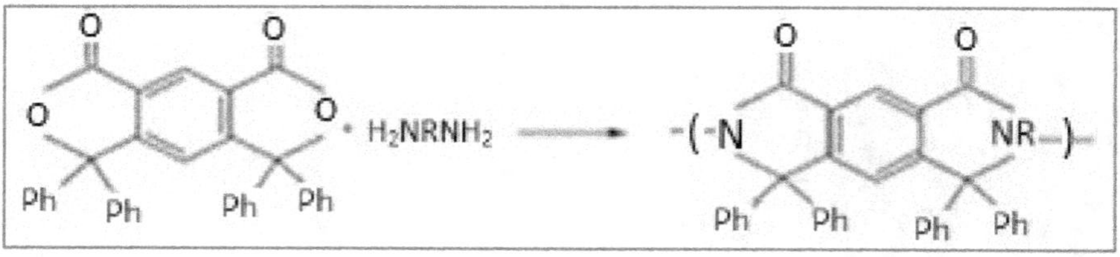

## c. Quinoline

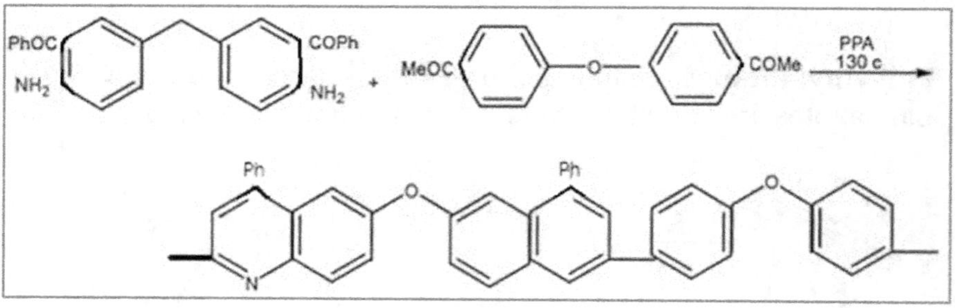

## d. Anthrazoline

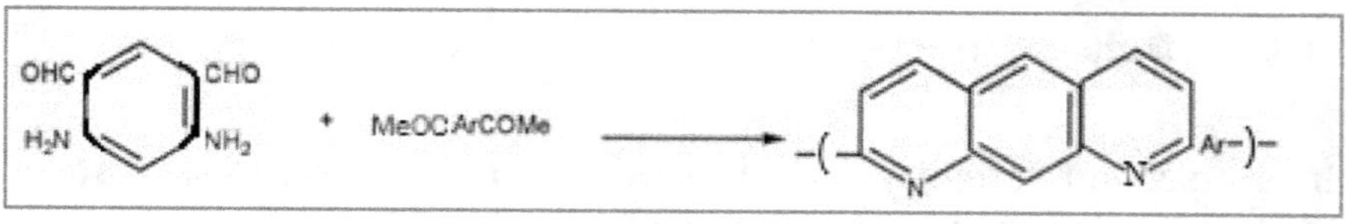

The third category comprises of **heterocyclic polymers** formed by polymer modification reactions and their main advantage is that the degree of reaction and hence concentration of product can be controlled and varied without changing the degree of polymerization. For example, **Imidazoline** has been used as a side chain group by reaction of an acrylonitrile-containing polymer (A) with ethylenediamine (B). The **imidazoline functional polymer** (C) finds extensive use as flocculation aids for the dewatering of sludge.

**Quinoxaline** is obtained by reaction of o-phenylenediamine with polyamide.

## 3.4 AS DYES AND PIGMENTS

Heterocyclic compounds have been used as colouring materials since ages like **Indigo**, **Haematin** etc.

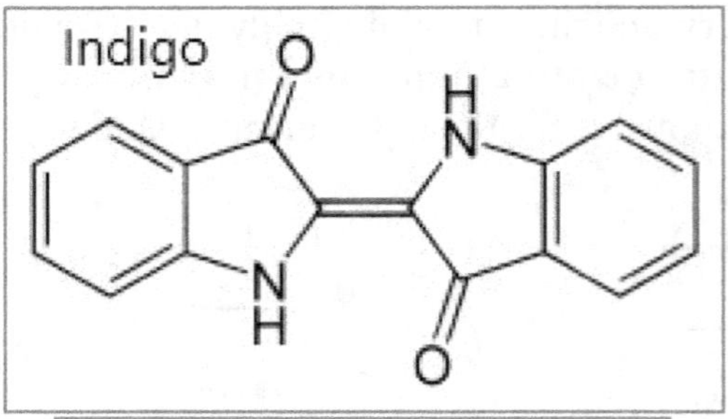

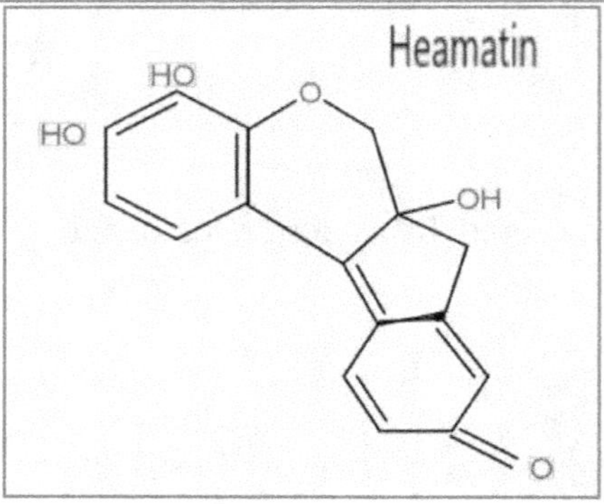

**Thio-indigo** is an important red dye, **Hydron pink FF**, a pink dye and **Anthranol Voilet,** formed by mixing indigo and thio-indigo, is of commercial significance.

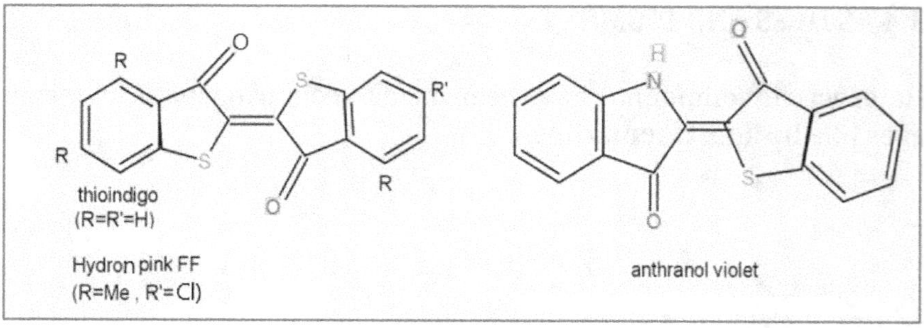

Several other five-membered heterocycles such as thiophene, thiazole and oxazole are annellated in anthraquinone series to yield the **Yellow Dye, Flavanthrone** (yellow), **Indanthrone** (blue) etc.

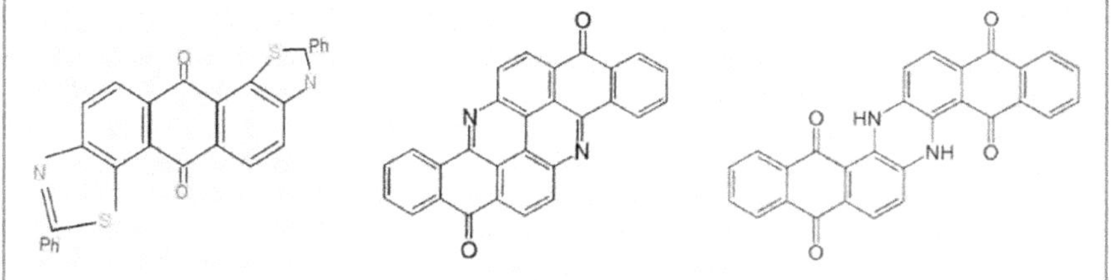

Yellow Dye  Flavanthrone  Indanthrone

The dyes of the class azine include oxazine, thiazine, acridine, and xanthones which are oldest of synthetic dyes with general structure:

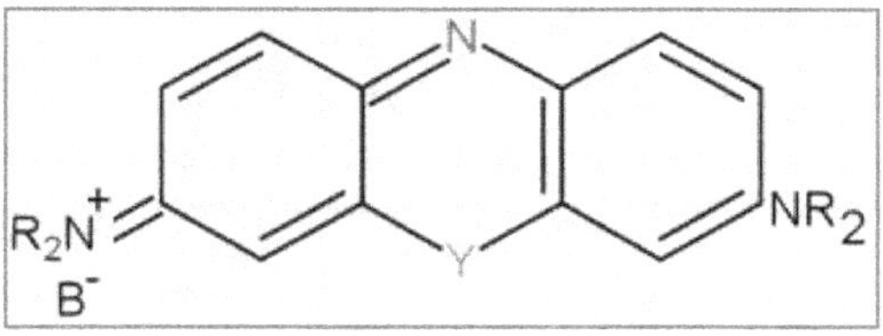

The colours provided range from violet to greenish blue as **Y** varies from N (azine) to O (oxazine) to S (thiazine) for eg- **Meldola's Blue.**

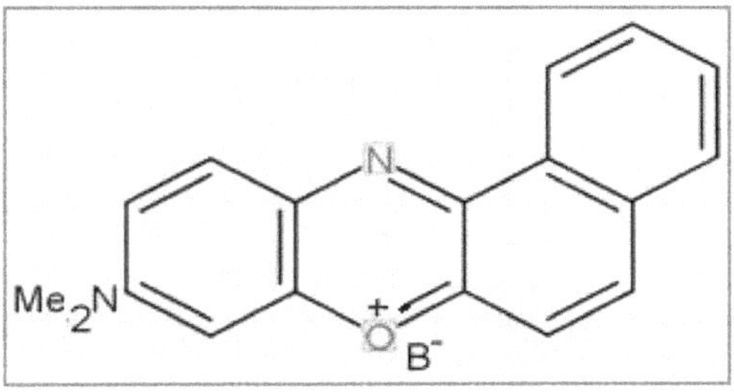

**2-Aminobenzothiazole** derivative is also a good example of azo dyes.

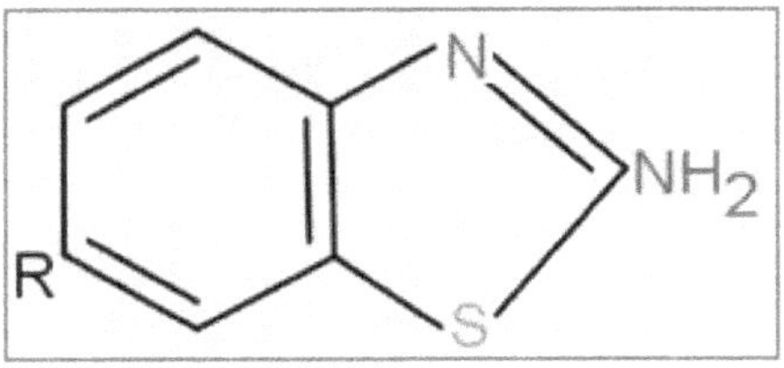

Azo pigments have been in use since long and greatly contribute to most of the organic pigment production. For eg- **1-phenyl-3-methyl-**

**5-pyrazolone derivative** (orange) and **yellow pigment** prepared by condensation of *p*-phenylenediamine with 3,3,4,5,6,7-hexachloroiso-indolinone are a few to mention.

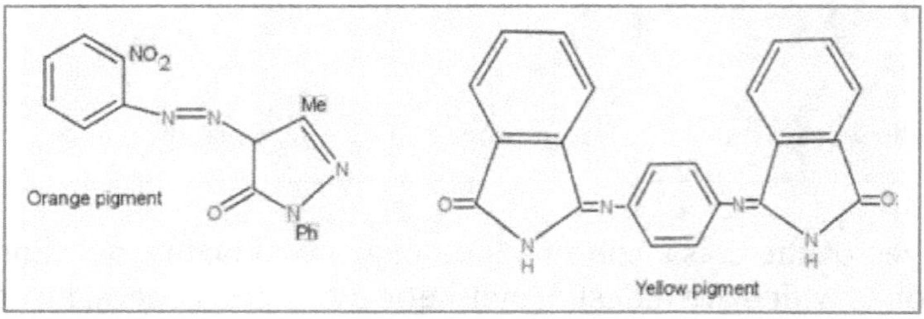

In the study by Gold, it has been stated that fluorescent brightening agents are colourless fluorescent dyes and they absorb a portion of UV radiation in sunlight and emit visible(blue) fluorescent light. The largest proportion of such agents is based on **Stilbene** and substituted etilbene for eg- one introduced by Geigy as **Tinopal RBS** for cotton and nylon.

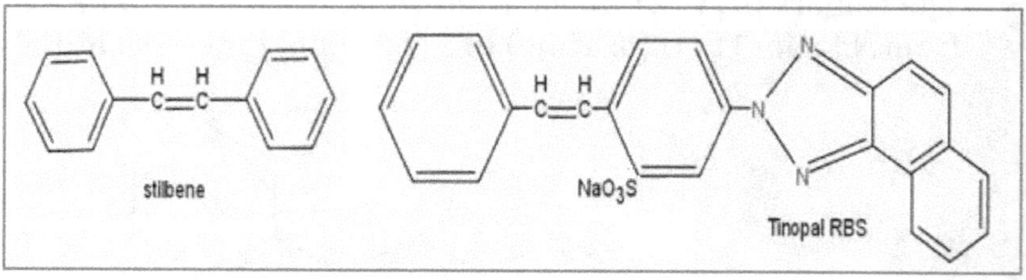

## 3.5 IN PHOTOGRAPHIC AND REPROGRAPHIC TECHNIQUES

The use of heterocyclic compounds in photographic and reprographic processes are in regulating the hardness of gelatine, regulating performance of silver halide emulsions, in absorption of unwanted radiation and as components in processing solutions. **Bis-aziridines** have an activated group (electron withdrawing substituents on nitrogen atom) which is important for stability of hardener.

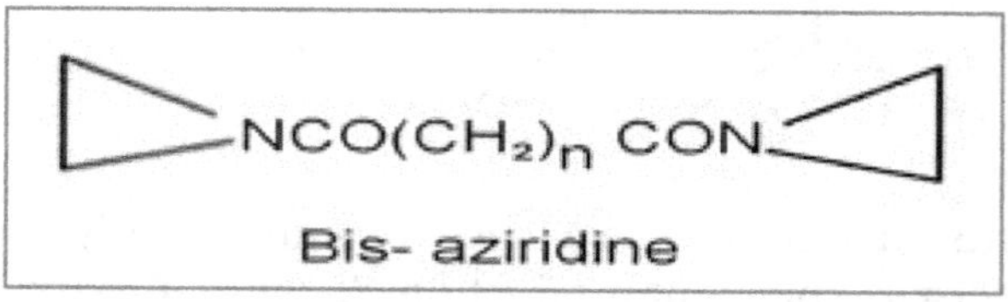

Compounds like **Piperazinedione** and **Benzimidazolinone** release hardeners (mostly formaldehyde) when required to prevent after hardening.

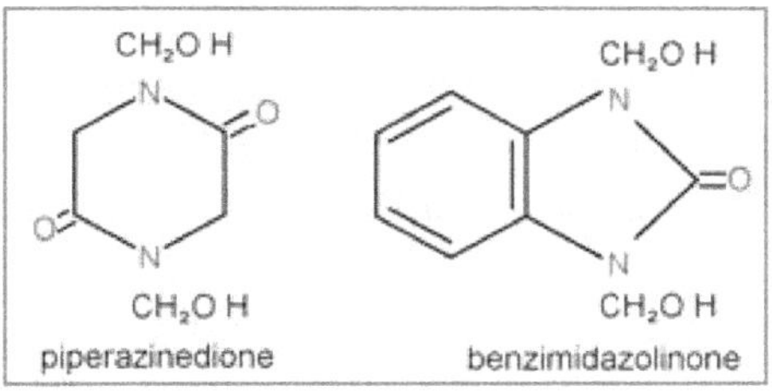

The sensitivity of silver halide is for ultra-violent and blue light only but the objects to be photographed are of all colours hence certain heterocyclic compounds are added in small amount to extend the sensitivity beyond blue. For e.g., the **dyes A and B** are sensitizing agents for red and infrared.

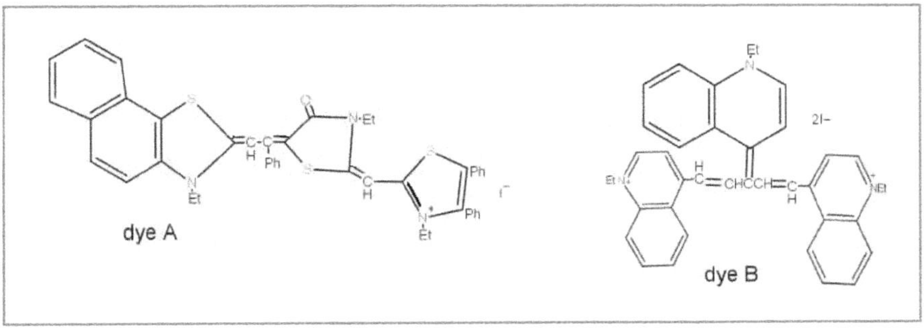

Compounds like **Amino-pyrazoline**, **Pyrazole** and **Pyrimidine** are useful as colour developing agents.

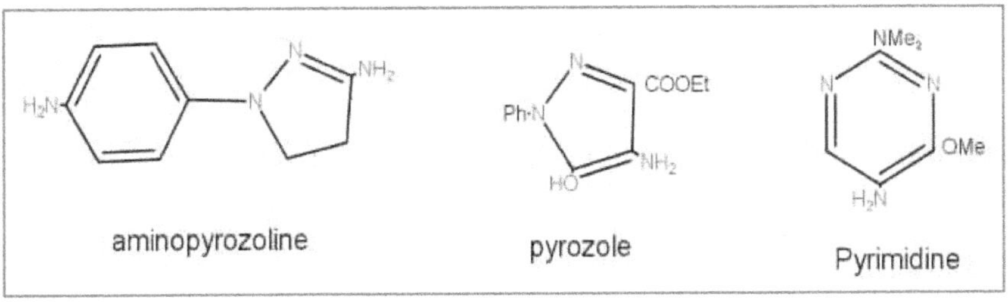

## 3.6 AS ADDITIVES

Additives are used for a variety of purposes as to improve processing (lubricants, emulsifiers, catalyst, vulcanizer), modifications of properties (plasticizers, antistatic agents, adhesion promoter, corrosion inhibitor, surfactant, colorant, optical brightener), to retard aging (antioxidant, antimicrobial agents) to reduce cost (filler, dilutant) and various others (deodorant, antifoaming agent, thickener).

Micro-organisms tend to act on nearly all materials under appropriate conditions. Most biocides are active inside the cell which is most likely site for metabolic activities. Biocides, in most of the cases, only inhibit the micro-organisms instead of destroying them which could have been a more appropriate solution.

| S. No. | COMPOUND AS BIOCIDE | USE |
|---|---|---|
| 1. | 1,2-Benzisothiazolin-3-one | Preservative for Adhesive |
| 2. | 2-Methyl-4-isothiazolin-3-one | Preservative for cosmetics, personal care products, cutting fluids |
| 3. | 2-Mercaptobenzothiazole | Fungicide, bactericide for cotton, rayon, adhesive |
| 4. | Hexahydro-1,3,5-triethyl-5-triazine | Preservative for adhesives, paints, rubber latex |
| 5. | N-(Tri-chloro-methyl-thio)-4-cyclohexene-1,2-dicarboximide | Preservatives for cosmetics, paints, plasticizers |
| 6. | Copper-8-hydroxyquinolinate | Fungicide for wood, plastics, adhesives, paints, fabrics |
| 7. | Alkyl-iso-quinolinium bromide | Preservatives in water treatment |

The compounds containing both sulphur and nitrogen, form the largest classes of heterocyclic biocides. Thus, **Thia-diazine, Iso-thiazoline, Benz-isothiazoline, Benzimidazole** and **2-Mercapto-Benzothiazole** are to name a few.

In the same way, compounds like **tris(2-hydroxyethyl) triazine** have proved to be an effective bactericide in cutting oils and coolants and **triethyl triazine** as preservatives for pigment slurries. **Dicarboximide**, in purified form, serves as a fungicide for cosmetic preparation while **Dicarboximide** and **Phthalimide** impart protection for plastics.



Sulphur, as the oldest colouring agent, is now only used along with other reagents. Many compounds are now used as sulphur donors like di-thiamine (**4,4-dithiomorpholine** and **4-morpholinyl-2-benzothiazole disulphide**).



Food preservatives like ascorbic acid and some of its derivatives that are used in food and **quinoline derivatives** are used to preserve colour of paprika and chilli powder.

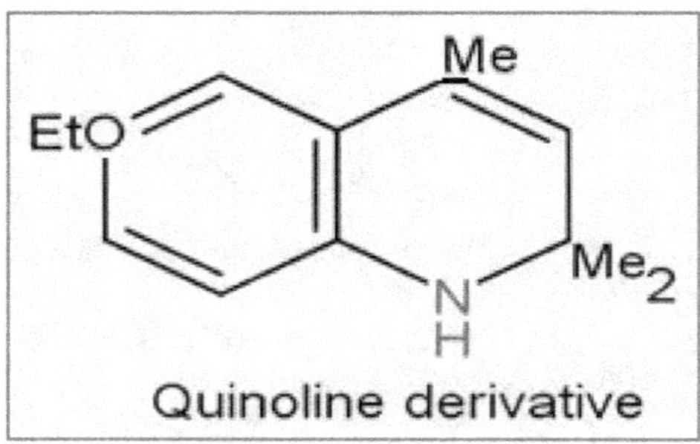


Heterocyclic compounds apart from the above usage are also widely used as veterinary products, conductors and in synthesis of non- heterocyclic compounds.


## REFERENCES

Meiser, W., Buechel, K.H., Kramer, W. and Grewe, F. (1973) (Farbenfabriken Bayer A.G.); Ger. Pat.2 201 063 (Chem. Abstr., 1973, 79, 105 257).

Devenport, J. D., Hackler, R. E. and Taylor, H. M. (1969) (Eli Lilly and co.); Fr. Pat. 1 569 940, (Chem. Abstr., 1970, 72, 100, 745)

Geigy, J.R. (1954), A.G., Br. Pat. 713 278 (Chem. Abstr., 1956, 50, 1092)

Dixon, J.K., Du Breuil, S. and Boardway, N.C. (1959) (American Cyanamid Co.); U.S. pat., 2 918 468 (Chem. Abstr., 1960, 54, 9971)

Farbenfabriken Bayer A.G. (1966) Neth. Pat. 66 07 054 (Chem. Abstr. (1967), 66, 95, 085).

Metiver, J. (1967) (Rhone-Poulene S.A.); Fr. pat. 1 482 025 (Chem. Abstr., 1968, 68, 105, 185).

Geigy J. R. (1963), A.G., Fr. Pat. 1 335 755 (Chem Abstr., 1964, 60, 1764).

Sandler, S. and Karo, W. (1977); polymer synthesis; Academic Press, New York, vol.2, p.232.

Limburg, W.W. and Semar, D. A. (1975); U.S pat. 3 877 936 (Chem. Abstr., 1975, 82, 58, 909).

Cotter, R. J. and Metzner, M. (1972); Ring forming polymerization; Academic press, New York, vol B-2 p.72.

Cassidy, P. E. and Syrinek, A. (1976); J. Polym. Sci., Polym. Chem. Ed., 14, 1485.

Stille, J. K. (1982); Macromolecules, 14, 870.

Capozza, R. (1974); U.S. pat. 3. 826 787 (chem. abstr., 1975, 82, 31, 884).

Nagakuba, K., Akutsu, F., Kawamura, N. and Miura, M (1977), 807, 6414.

Pinckney, E. (1976), Am. Dyest. Reptr., 65, 36.

Allen, R. L. M. (1971); Colour chemistry Nelson, London.

Friendlander, P. (1906); Ber.,39, 1060.

Scholl, R. (1907); Ber., 40, 1691.

Venkataraman, K. (1952); 'the chemistry of synthetic Dyes', Academic Press, New York, Vol. 1, P. 608.

Gold, H. (1971); D. R. Boer; in 'the chemistry of synthetic Dyes', Ed. K. Venkatara man; Academic Press, New York, Vol. 5, P. 535.

Birr, E. J. and Walther, W. (1960); (VEB Filmpabrik Agfa wolfen), Ger. Pat, 1 081 169 (Chem. Abstr., 1961, 55, 17, 323)

Kaszuba, F. J. (1952); (General Aniline and film corp.), U.S. pat. 2 586, 168 (chem. Abstr. 1952, 46, 3889)

July, G., Knott, E. B. and Pollak, F. (1956); (Eastman Kodak co.), U.S pat. 2 732 316 (chem. Abstr., 1956, 50, 7640).

Saunders, D. G. (1956); J. Chem. Soc., 3232.

Hinton, A. J., Turner, J. N. and Morley, J. S. (1961); (ICI Ltd.), Br. Pat. 844 541 (chem. Abstr., 1962, 56, 10, 637). Technical Bulletin CS-471, 'Kathon 886F Industrial Microbicide; Rohm and Haas co., Philadelphia, PA, USA.

MeCutcheon's 1980 Functional Materials (1980), Manufacturing Confectioner Publishing co, Glen Rock.

Trotz, S. I. and Pitts, J. J (1981); in 'Encyclopedia of chemical technology'; Wiley- Interscience, New York, 3rd ed., vol. 13, P. 223.

Throdahl, M. C. and Beaver, D. J. (1945); Rubber chem. Technol., 18, 110.

# CHAPTER 4

## NOMENCLATURE OF HETEROCYCLIC COMPOUNDS

Heterocyclic compounds are derived from the carbocyclic compounds by the replacement of one or more carbon atoms by the heteroatom or heteroatoms like N, O, S, etc. Heterocyclic chemistry came into picture as early as 1818 when Brugnatelli isolated alloxan from uric acid and in 1832, Dobereiner produced furfural by treating starch with sulphuric acid.

alloxan
(1, 3-diaxinane-2, 4,5,6-tetrone)

furfural
(furan-2-carbal-dehyde)

In earlier days of organic chemistry, names were given to compounds as a means of identifying them, usually before their structures were known. But for the need of a systematic nomenclature three methods have become prevalent:

- **Common Nomenclature** – conveys little or no structural information but it is still widely used.
- **Hantzsch-Widman** (IUPAC /Systematic)- method designed so that one may deduce from it the structure of the compound.
- **Replacement Method**

## 4.1 COMMON NOMENCLATURE

Each compound is given the corresponding trivial name. This usually originates from the occurrence of the compound.

For compounds having more than one heteroatom of same type, numbering starts at saturated one.

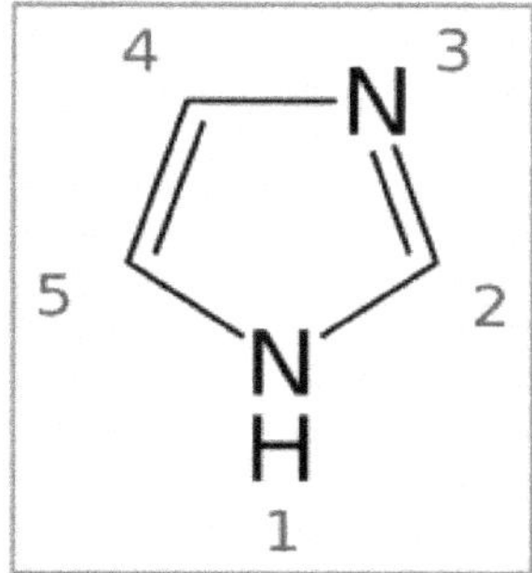

If there are heteroatoms of different types, number starts according to priority: **O > S > N** and continues in the direction so that another heteroatom gets the lower atom.

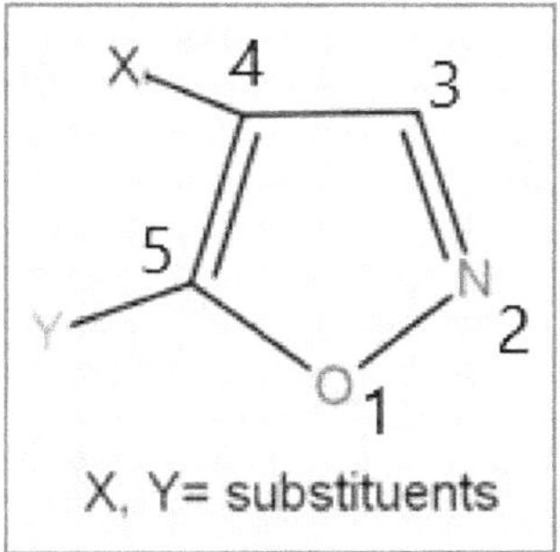

X, Y= substituents

If substituents are present, their positions are according to the atoms to which they are attached.

For two or more than two saturated atoms, as low as possible numbers are given.

32

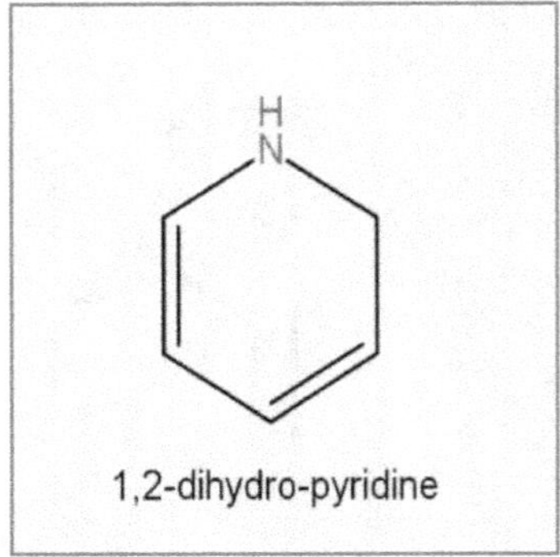

1,2-dihydro-pyridine

## TRIVIAL NAMES

These names were given before their structural identification based on characteristic properties or on the source from which they were obtained.

### Saturated Heterocycles

pyrrolidine        piperazine        piperidine        morpholine

### Five - membered heterocycles (one/two heteroatoms)

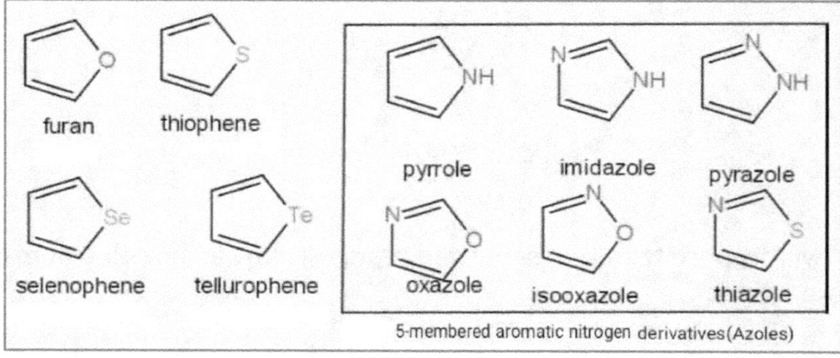

<u>Six membered heterocycles (one/two heteroatoms)</u>

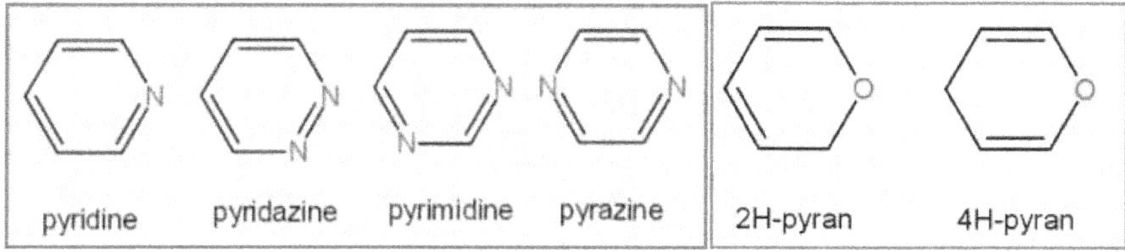



## Fused Heterocycles

A) Fused Azoles:

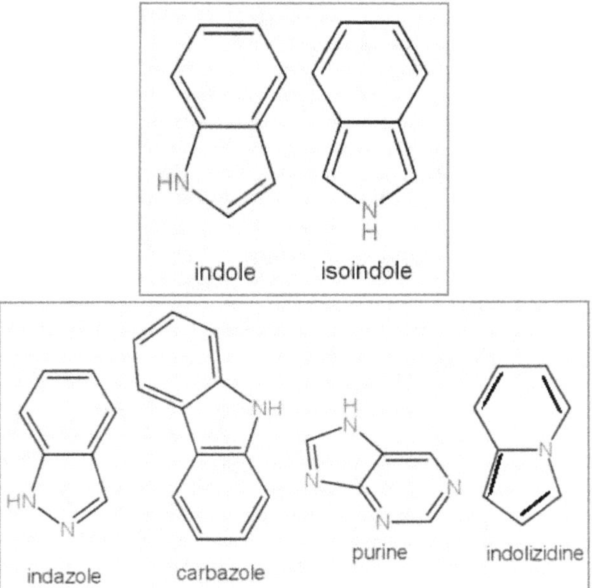

34

B) Fused Azines:

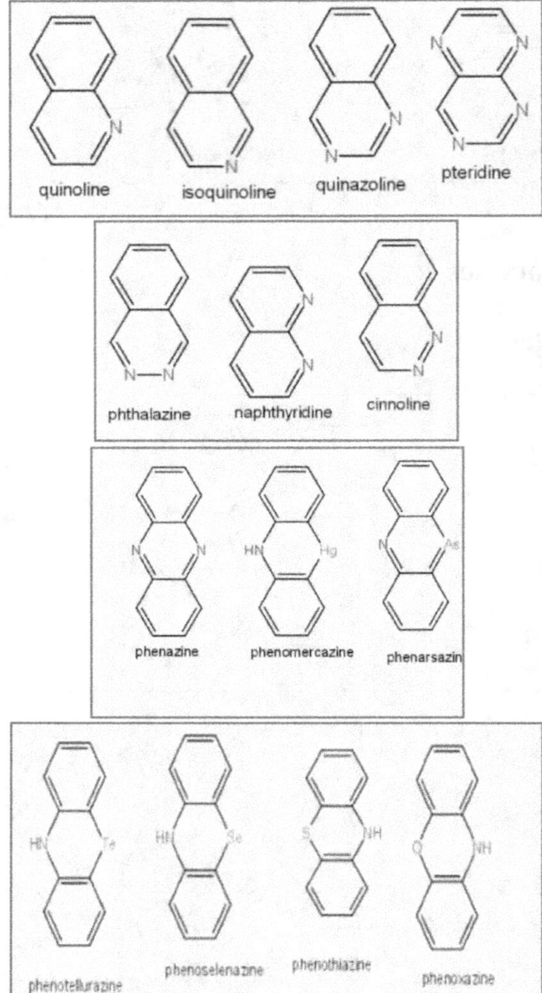

# C) Others:

chromen4one
chrom en-2-one
flavone
arsindole
phosphindole
arsinoline
phosphinoline
phenomereurin
acridasine
arsanthrene
arsanthridine
phosphanthrene
selenanthrene
thianthrene
phenothiarsine
phenoxantimonine
phenoxaphosphine
phenoxatellurine
phenoxaselenine

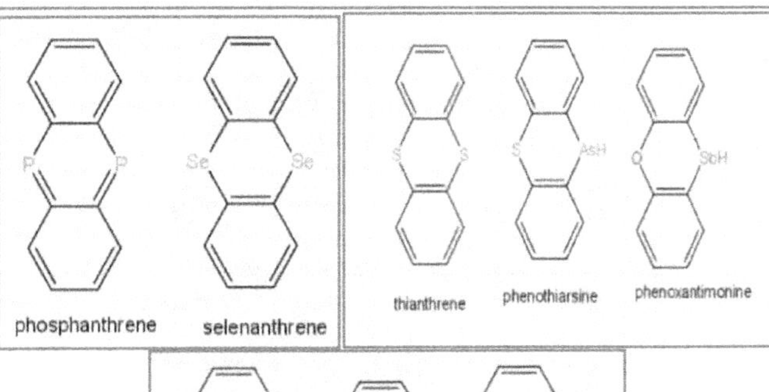

## 4.2 HANTZSCH-WIDMAN NOMENCLATURE

Hantzsch - Widman nomenclature was proposed by German chemists **Arthur Hantzsch** and **Oskar Widman** in 1887-1888. This method is used for naming three to ten membered heterocyclic compounds. This nomenclature specifies the ring size, nature, type and position of heteroatoms. A. Hantzsch and O. Widman, who propounded this system of nomenclature used the principle for naming by combining prefix (indicating number and nature of heteroatoms) with stems (indicating ring size and its saturation or unsaturation). Table 1 shows the Non-metallic elements covered by Hantzsch Widman system along with their prefixes, all of them ending in 'a'.

Table 1 Hantzsch – Widman system: prefixes for heteroatoms

| Element | Valency | Prefix |
| --- | --- | --- |
| Fluorine | I | fluora |
| Chlorine | I | chlora |
| Bromine | I | broma |
| Iodine | I | ioda |
| Oxygen | II | oxa |
| Sulphur | II | thia |
| Selenium | II | selena |
| Tellurium | II | tellura |
| Nitrogen | III | aza |
| Phosphorus | III | phospha |
| Arsenic | III | arsa |
| Bismuth | III | bisma |

| | | |
|---|---|---|
| Silicon | IV | sila |
| Lead | IV | plumba |
| Boron | II | bora |
| Mercury | II | mercura |

If the heteroatoms are more than one, then IUPAC prefixes are used; the final 'a' is dropped if the prefix is followed by a vowel. For example, tetra + oxepane will be tetroxepane instead of tetraoxepane. Two or more different heteroatoms of same type are represented by di-, tri-, etc. However, two or more heteroatoms of different kind are prefixed in the order listed in table 1, the stem suffixes are used for the size of ring and are indicated in table 2.

**Table 2 Suffixes for stems**

| RING SIZE | SATURATED | UNSATURATED |
|---|---|---|
| 3:　with N atom. | - iridine | -irine |
| 3:　without N atom. | -irane | -Irene |
| 4:　with N atom | -etidine | -ete |
| 4:　without N atom | -etane | -ete |
| 5:　with N atom | -olidine | -ole |
| 5:　without N atom | -olane | -ole |

In case of ring containing more than one heteroatom; group 7th elements (halogens) are provided the highest priority and descending down the column, so on moving to next lower group. The following examples will clarify the mentioned point.

## 4.3 REPLACEMENT NOMENCLATURE

Replacement system of nomenclature uses most of the names of Hantzsch - Widman system leading to differently placing of locants for example,

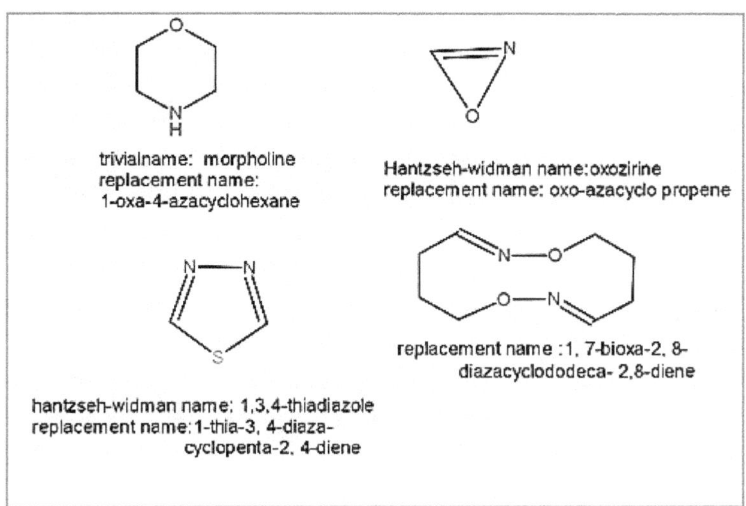


Rings that contain of elements other than carbon weather in **homogeneity** or in repeating units are named in the same way as Hantzsch - Widman system in replacement system. In rings composed of entirely one element, names for chains of like atoms prefixed with **"cyclo-"** Is used.



For fused heterocyclic compounds, replacement names may be adopted as:

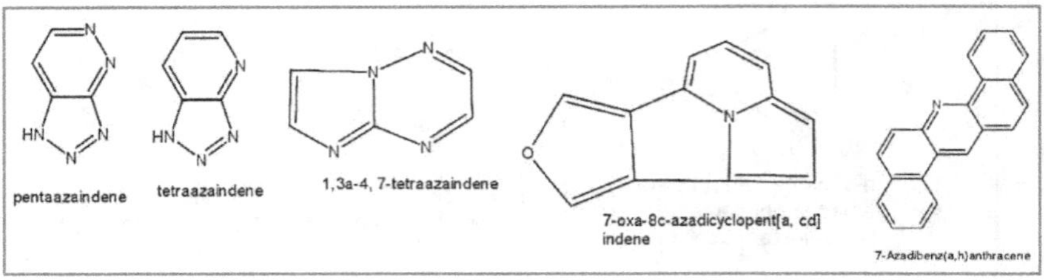

40



## 4.4 FUSION NAME SYSTEM

Fusion systems refers to the combination of two or more ring units in which each component has minimum of one bond in ortho-fused or ortho- and peri- fused.

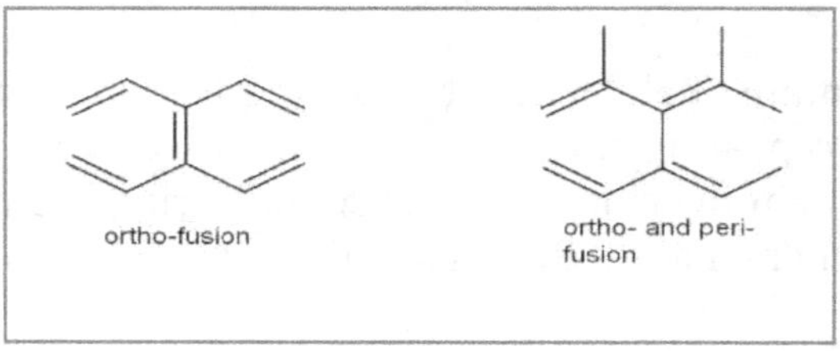

Fusion nomenclature is formed by adding prefix of one component to the trivial name of base component and thereby provides names to compounds with maximum number of non-cumulative double bonds. Base component should be heterocyclic (in order of preference as in table 1 with an exception to nitrogen which has the highest priority) and should contain more of the possible rings. When core compound name ends with 'e' then this letter is replaced by 'o' to the trivial/Hantzsch-Widman name.

| TRIVIAL/HANTZSCH-WIDMAN NAME | FUSION NAME |
|---|---|
| Acenaphthylene. | Acenaphtho- |
| Benzene. | Benzo- |
| Furan. | Furo- |
| Imidazole. | Imidazo- |
| Naphthalene. | Naphtho- |
| Pyridine. | Pyrido- |
| Pyrimidine. | Pyrimido- |

For monocyclic hydrocarbon core compounds, prefixes to be used are cyclopropa-, cyclobuta-, etc. ending in –ene showing the maximum number of non-cumulative double bonds. For example,

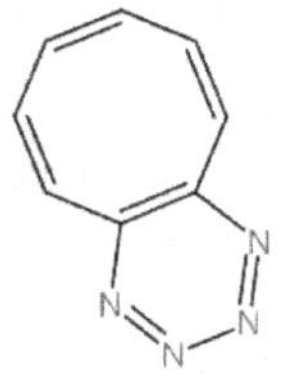

cyclo-octatetrazine

The position of fusion is shown by the numbering of core compound stating from bond 1,2 with alphabets in lowercase and italics in square brackets. For example.

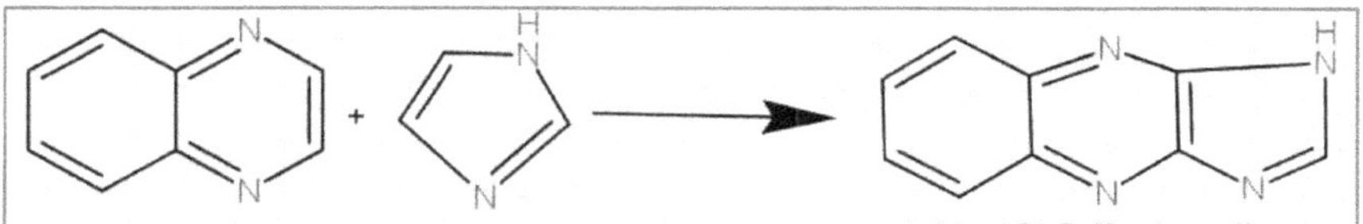

imidazo[4,5-6]quinoxaline

Original numbering changes after fusion and new numbers are assigned to the compound according to the correct orientation of the structure and then applying the set of priority rules. In the correct orientation method, the full structure is arranged in such a way that maximum number of rings are in horizontal row and minimum number is below and to the left of it.

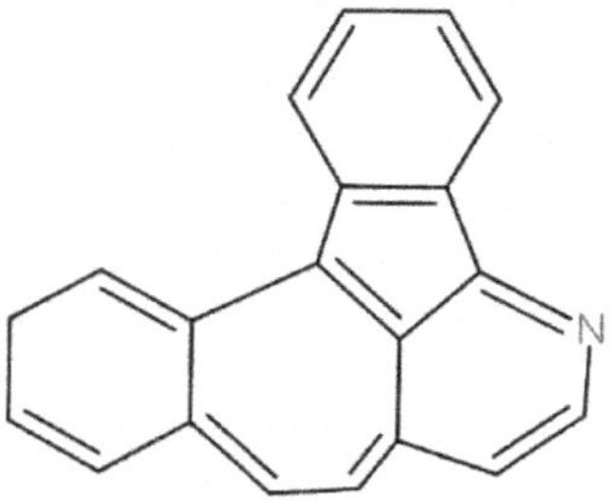

If there is carbon atom at the fusion of two rings, then it is not given a separate number but a **lowercase** alphabet after the number of atoms immediately preceding it. However, if heteroatom is common to two rings, then it receives an individual number.

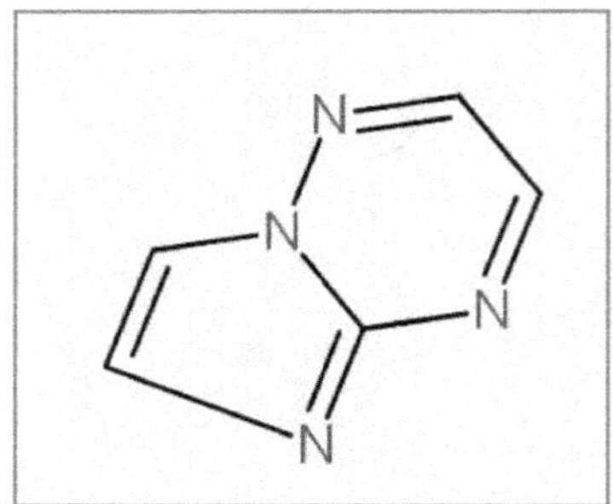

If the heterocyclic compound is **benzo – fused**, then the name starts with a number of heteroatom and then the trivial or Hantzsch - Widman name follows.



## 4.5 SPIRO NAMES

Spiro-linked compounds are those in which there is no ring fusion or bridge present. Such compounds are named as **"spiro [x, y] alkane"**, where x, y are numbers of carbon atoms including the spiro atom. however, spiro atom/atoms is /are given the lowest number. the spiro compound is named by Replacement Nomenclature and numbering begins with the atom next to spiro atom for example,

5-oxa-8-thiaspiro[3.4] octane

6,15-dithiadispiro[4.1.5.3] pentadecane

## DIFFERENT VALENCE STATE OF HETEROATOM

The elementary systems of nomenclature are applicable to those compounds which hold standard valency for heteroatom viz., 3 for nitrogen group,2 for sulphur group and 1 for halogens. In the recent naming system of such compounds, $\lambda$ symbol is used for heteroatom (suggesting that it is a Ligand) and valency is given as numeral super-script for example,

$1,3\lambda^5$ - Thiarsole

$7\lambda^4$ - [1,2] Dithiolo-[1,5-b] [1,2]dithiole

$5\lambda^5$ - phosphospiro [4,4]nonane

## NAMES WITH RESPECT TO INDICATED HYDROGEN

The unsaturation of cycloalkanes shows that for fully unsaturated skeletons and addition of hydrogen across double bond is indicated by numbering.

44

5,6,7,8,9,10,11,12,13,14-
Decahydro-6-azabenzo-
cyclododecene.

6-Azobenzo-cyclododecene

However, in the indicated hydrogen approach, hydrogen is located by its position with an italicised capital H in continuation.

2*H*-pyran　　　4*H*-pyran　　　2(3*H*)-pyridone　　1*H*-1,2-Thiazine

1*H*-azirine　　2*H*-azirine　　2(1*H*)-pyridone　　2H-1,2-Thiazine

## 4.6 NAMING OF BRIDGED FUSED STRUCTURES

Bridged structures are often treated as polycyclic structures and were earlier cited as divalent radical substituent which were numbered independently irrespective of the remaining skeleton, for example, **3,7-Ethylenethieno[3,2-c] pyridazine**.

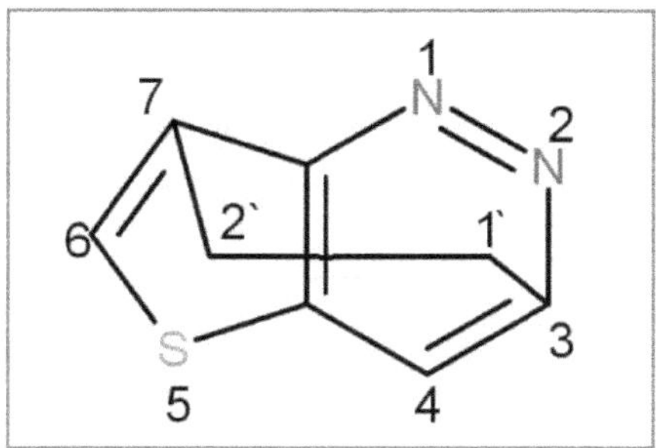



But now the bridge is a part of the ring system and is numbered in continuation with the skeleton. Some of the special prefixes used for bridge are:

| GROUPS | Names |
| --- | --- |
| -CH< | Methene |
| >C< | Methyno |
| -CH$_2$-CH$_2$- | Ethano |
| -CH=CH- | Etheno |
| -CH$_2$-CH$_2$-CH$_2$-CH$_2$- | butano |
| -CH$_2$-CH=CH-CH$_2$- | But[2]eno |
| -C$_6$H$_4$- | Benzeno |
| -N=N-NH- | Azimino |
| -N=N- | Azo |
| -NH-NH- | Biimino |
| -NH- | Imino |
| -NH-CH$_2$-CH$_2$- | Iminoethano |
| -O-O- | Epidioxy |
| -S-S- | Epidithio |
| -S- | Epithio |

| -S-O- NH- | Epithioximino |
|---|---|
| -O-NH- | Epoxyimino |
| -O-CH$_2$- | Epoxymethano |
| -O-N= | Epoxynitrilo |
| -O-S- | Epoxythio |
| -N= | Nitrilo |

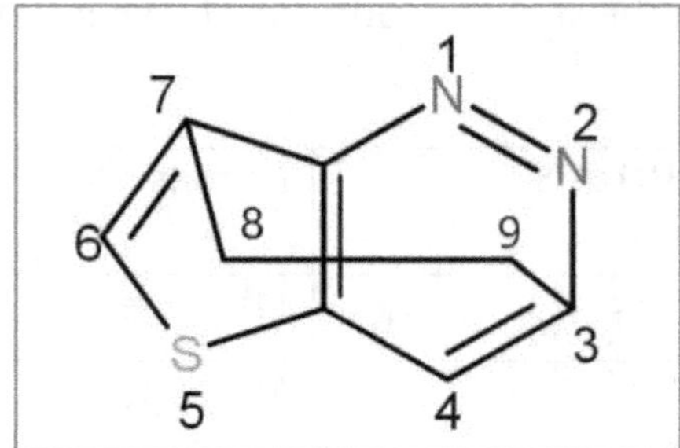

**3,7-Ethanothieno[3,2-c] pyridazine**

## 4.7 VON BAEYER SYSTEM

This system was evolved for naming bicyclic bridged saturated compounds which later found application in polycyclic system. The rules of Von Baeyer system are:

- A bridge head is any skeletal atom or a ring system which is bonded to three or more skeletal atoms.
- The main ring is selected which includes as many skeletal atoms of the compound as possible.

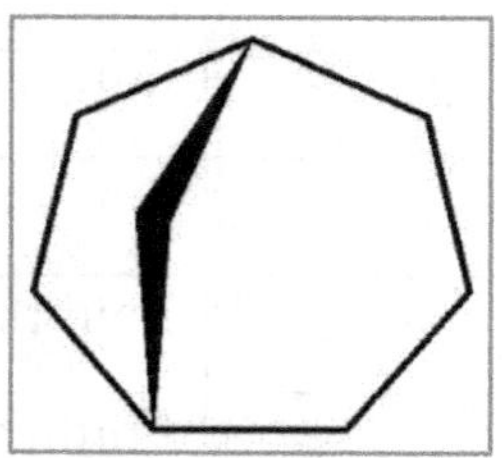

Bicyclo [3.2.1] octane

- A bicyclic system, having a main ring and main bridge, contains the term bicyclo-, tricyclo-, tetracyclo- etc indicating the number of rings.
- Bridge length is indicated by numbers separated by full stops and placed in square brackets. The numbers being in decreasing order of Size.
- The atoms are numbered from bridgehead atoms via the longest path to second bridgehead atom while numbering continues round the main ring.

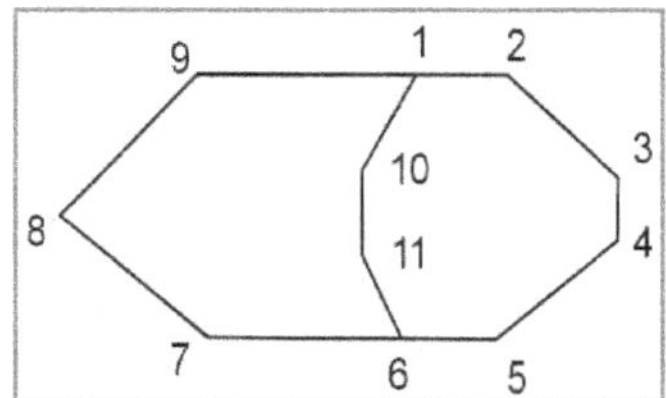

- If there are more than one bridge, the main bridge is selected to include as many as possible of the atoms not in the main ring.

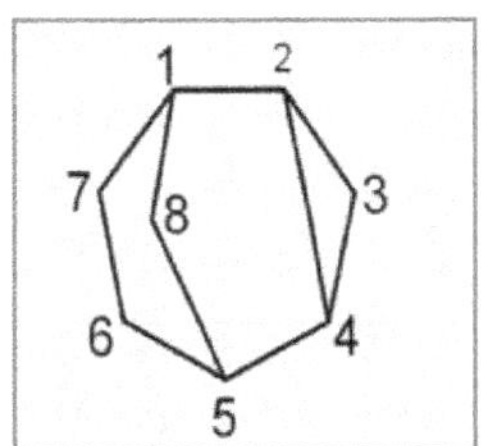

tricyclo[3.2.1.0$^{2,4}$] octane

## Examples:

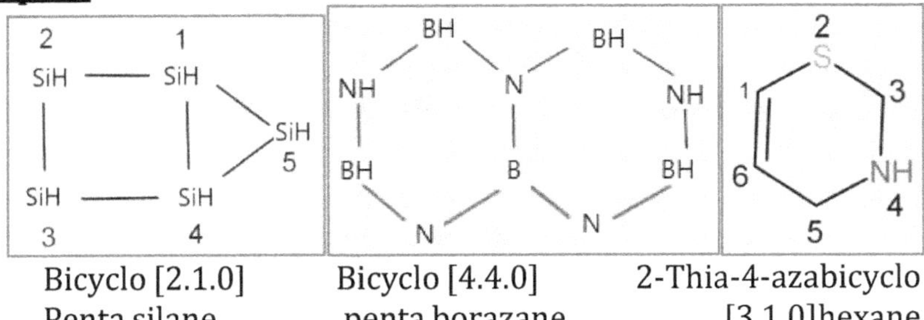

Bicyclo [2.1.0]
Penta silane

Bicyclo [4.4.0]
penta borazane

2-Thia-4-azabicyclo
[3.1.0]hexane

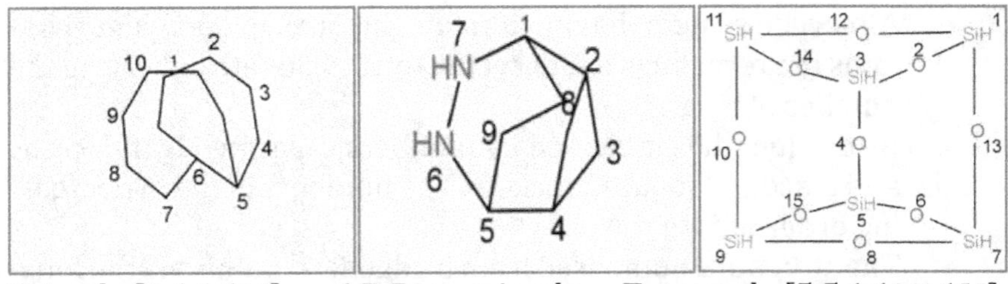

Tricyclo [4.4.1.1$^{1.5}$] dodecane     6,7-Diazatricyclo [3.2.2.0$^{2.4}$] nonane     Tetracyclo [5.5.1.1$^{3.11}$.1$^{5.9}$]- hexa siloxane

## 4.8 NAMING OF NATURAL HETEROCYCLIC COMPOUNDS

Heterocyclic components are present in many classes of natural products. The common system of naming the compound is the **trivial system** where little or no structural information is given and the names are based on the Latin names of the species form which the compound was isolated. For example,



Other compounds have been named in a way giving detailed information about the structure.

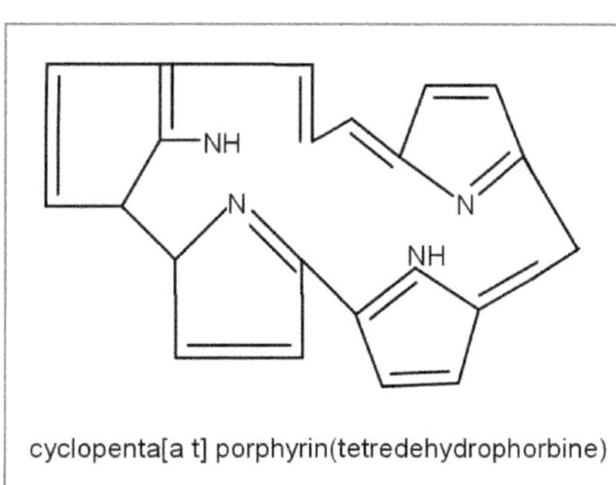

cyclopenta[a t] porphyrin(tetredehydrophorbine)

α-D-ido-hexapyranose

Tetrabenzo[b,g,l,q][5,10,15,20]tetraazaporphyrin

(Phthalocyanine)

α-D-galacto- hexopyranose (or α-D-

galactopyranose)

4,6-dideoxy-5',6'-dihydro- α-D- ido-

hexopyranose[6,5,4-de] [,1,3] oxazine

## 4.9 NAMING OF MISCELLANEOUS COMPOUNDS

The products formed as a result of substitution on the parent compound, are named by either prefixing or suffixing the substituent related to the name of the parent structure. For example,

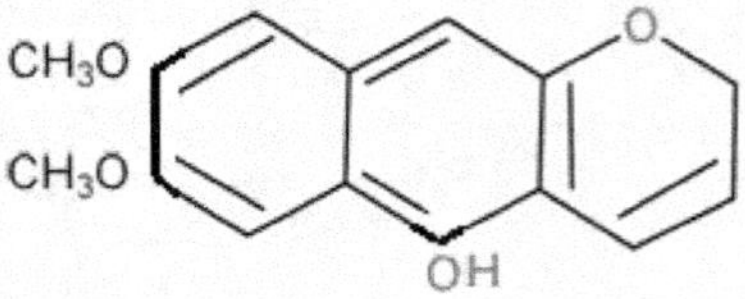

7,8-Dimethoxy-2H-naphtho-[2,3-b]
pyran-5-ol

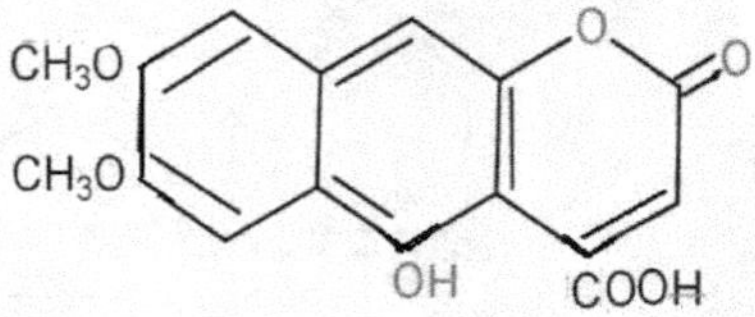

5-hydroxy-7,8-Dimethoxy-2-oxonaphtho-[2,3-b]
pyran-4-carboxylic acid

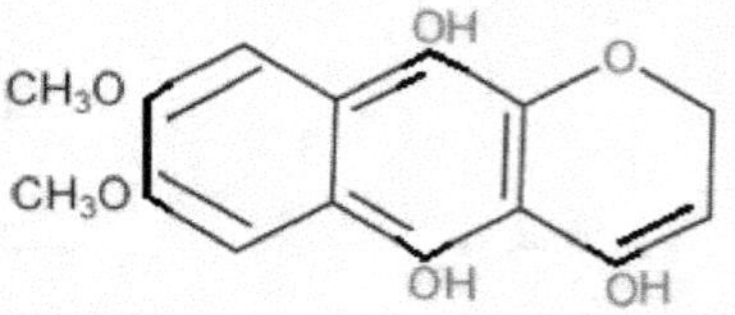

7,8-Dimethoxy-2H-naptho[2,3-b]-
pyran-4,5,10-triol

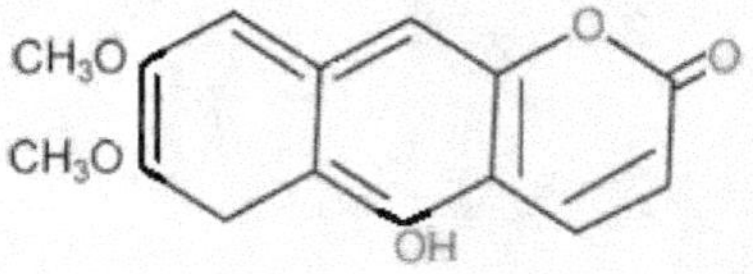

5- hydorxy-7,8-dimethoxy napththo
[2,3-b]pyran-2-one

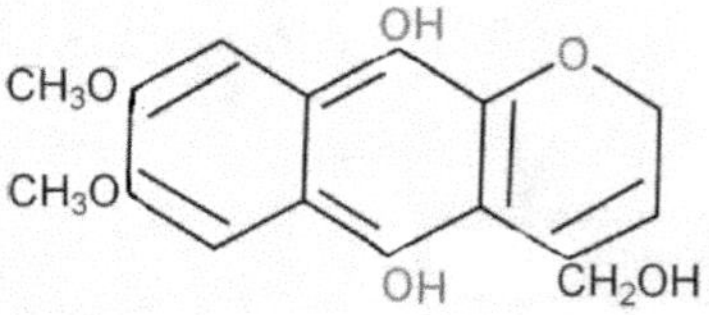

4-hydroxymethyl-7-8-dimethoxy-2H-naphtho-[2,3-b]pyran-5,10-diol

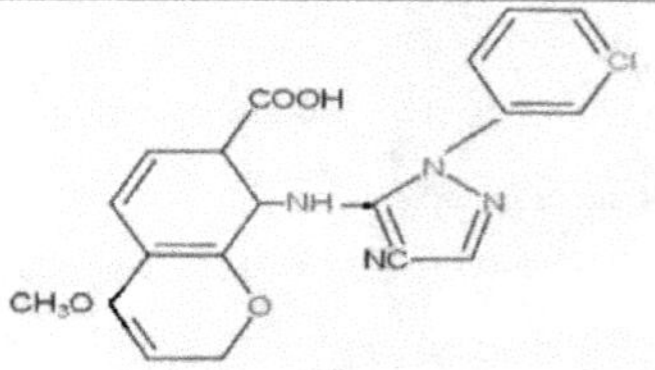

8-[1-(2-chlorophenyl)-4-cyanoimidazol-5- ylamino]-4-
methoxy-2H-chromene-7-carboxylic acid

## 4.10 NAMING OF HETEROCYCLIC RINGS CONTAINING PHOSPHORUS

The nomenclature of phosphorus compounds is quite difficult. However, some common names may be as under:

| S.No. | Ring Size | Single Bonded | Double/Triple Bonded |
|---|---|---|---|
| 1 | 3-membered | Phosphirane | Phosphirene |
| 2 | 4-membered | Phosphetane | Phosphate |
| 3 | 5-membered | Phospholane | Phosphole |
| 4 | 6-membered | Phosphorinane | Phosphorin |
| 5 | 7-membered | Phosphepane | Phosphepin |
| 6 | 8-membered | Phosphocane | Phosphocin |
| 7 | 9-membered | Phosphonane | Phosphonin |
| 8 | 10-membered | Phosphecane | Phosphecin |

For example:



## 4.11 NOMENCLATURE OF HETEROCYCLIC RINGS CONTAINING BORON

Boron containing heterocyclic compounds have been assigned the name "**Borozaro**". Heterocyclic compounds contain heteroatom replaced as a replacement of carbon atoms in the parent aromatic compound and hence the name of the ring is after the parent ring prefixed to indicate heteroatom. 'Aro' prefix is added after heteroatom "Bora" for example: 2,1-Borazarobenzene is named as 2,1-borazarene.

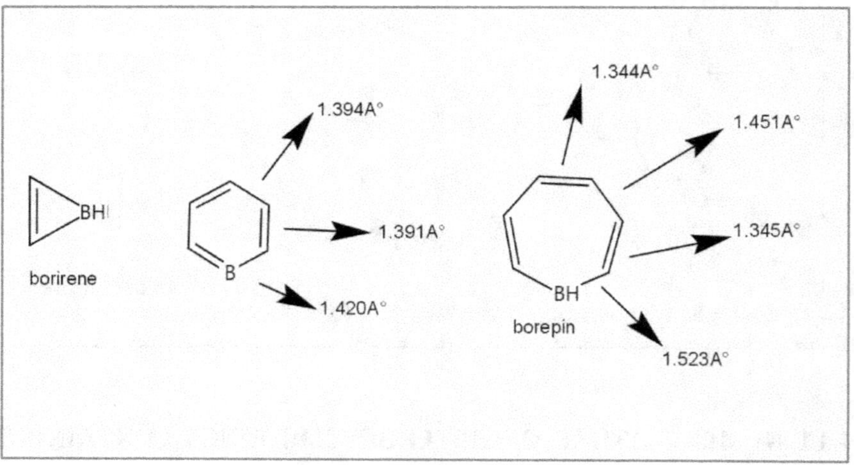


## REFERENCES


Fletcher, H. J., Dermer, O. C. and Fox, R. B. (1974); Nomenclature of organic compounds; Adv. chem. ser., 126, 49.
Hantzsch, A. and Weber, J. H. (1887); Ber., **20**, 3119.
Widman, O. (1888); J. Prakt. chem., **2,** 38, 185.
Sewar, M. J. S. (1964); Prog. Boron chem., **1**, 235.

# CHAPTER 5

## THREE & FOUR MEMBERED HETEROCYLIC COMPOUNDS

### 5.1 INTRODUCTION

Three – membered heterocyclic compounds may have one or two heteroatoms. Accordingly, such ring systems have Azirides, Oxiranes, Thiiranes, Azirines, Oxirenes, Thiirenes, Diaziridines, Diazirines, Ox - aziridines as members. For members having one heteroatom, one carbon atom of cyclopropane is replaced by either nitrogen (**AZIRI-DINE**), oxygen (**OXIRANE**) or sulphur (**THIIRANE**) or. This change brings about change in physical and chemical properties due to ring strain.

Heterocyclic derivatives of cyclo-butane formed by replacing a methylene group($-CH_2$) with heteroatom falls into the category of four-membered heterocyclic compounds. If the heteroatom is nitrogen, the four-membered heterocycle is **AZETIDINE** and **OXETANE** and **THIETANE** for respectively oxygen and sulphur as heteroatoms. Four-membered heterocyclic compounds are more stable than three-membered heterocycles as their rings are less strained.

### 5.2 3 - MEMBERED HETEROCYLIC COMPOUNDS

### 5.2.1 AZIRIDINE: $C_2H_5N$

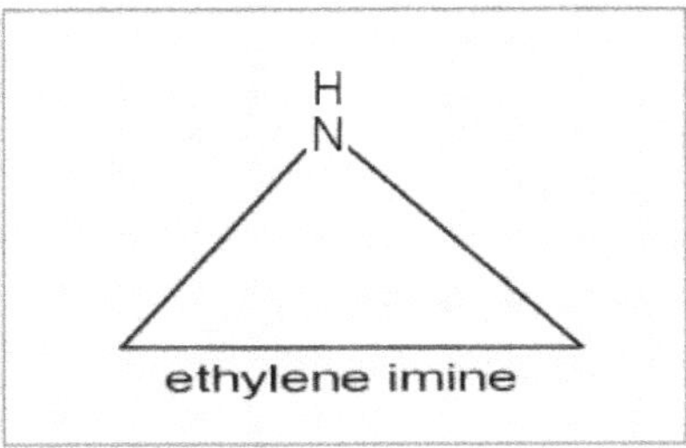


### SYNTHESIS OF AZIRIDINE

(a)     When β -Bromo ethylamine is heated; it converts into a cyclic structure in the presence of ethanolic potassium hydroxide.

54

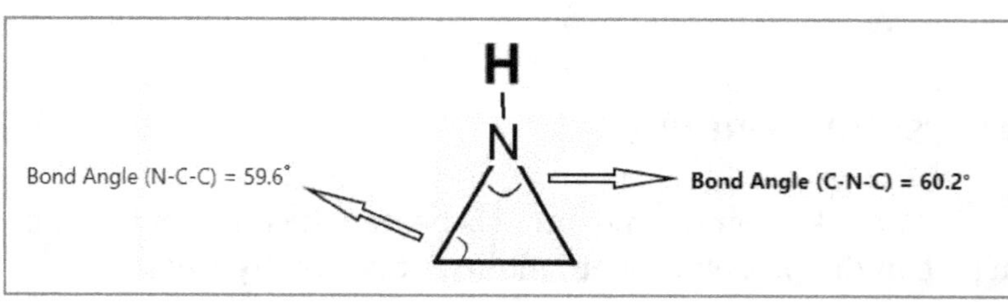

(b)      Aziridines can also be prepared when oxiranes are heated with substituted imino-phosphorane at 140°C. The important condition for this reaction to occur is absence of moisture and solvent. By this reaction **1-methyl-2-phenylaziridine** is obtained.

(c) **Cycloaddition** of azides with olefins under high pressure gives aziridines.

## PROPERTIES OF AZIRIDINES

a) Aziridines are clear colorless oily liquids with ammonia like odour.
b) They are miscible with water.
c) boiling point = 56°C; melting point = - 77.9°C
d) The C-C bond length in aziridines is 1.48 Å. The internal bond angles of aziridines are as follows:

e) Aziridines lead to the ring opening products in the presence of mineral acids, acetyl chloride and water.

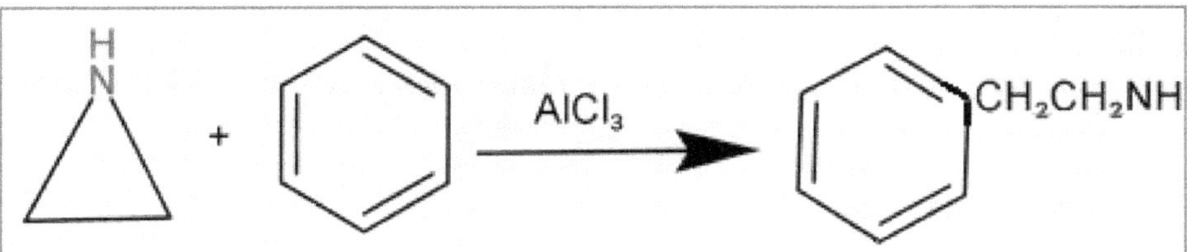

f) In the presence of aluminum chloride, aziridines react with benzene to form β- phenylethylamine. (**Friedel- Crafts Reaction**).

g) Hydrogen atom present at nitrogen is replaced by the following means-

## APPLICATIONS OF AZIRIDINES

Aziridines are popular as biological alkylating and anti-cancer agents. The naturally occurring **Mitomycin C** has been reported for antibiotic and anti-tumor activity. Aziridines are irritants of mucosal surfaces including eyes, nose and skin.

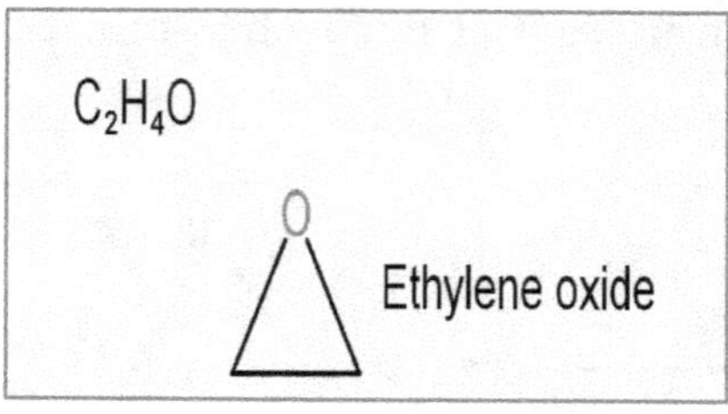

**MITOMYCIN C**

Aziridines are also used as components of reactive dyes for cellulose fibers.

## 5.2.2 OXIRANES



## SYNTHESIS OF OXIRANES

a)    Aldehydes when react with diazomethane yield cyclic oxide with or without ketone.

$$CH_3CHO \xrightarrow[-N_2]{CH_2N_2} CH_3CH{-}CH_2 + CH_3OCH_3$$

b)   The simplest method of preparation is the **direct oxidation** of ethylene by air over a silver catalyst at high temperatures.

$$H_2C{=}CH_2 \xrightarrow[\text{high temp}]{Ag,[O]}$$

The double bond breaks and leads to a cyclic structure. Alternatively, substituted alkenes yield same result with hydrogen peroxide or organic hydroperoxides and metal catalysts.

[2-acetoxy-2,3-diethyl oxirane]

## PROPERTIES OF OXIRANES

a) Oxiranes are colorless liquid.
b) Boiling Point = 10.7°C; Melting Point = -111.3°C.
c) They are miscible with water.

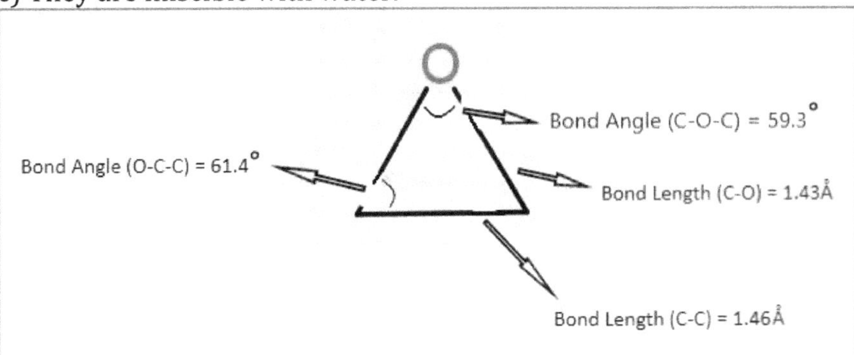

58

d) Oxiranes are **reduced** by metal hydrides forming alcohols with the same number of carbon atoms.

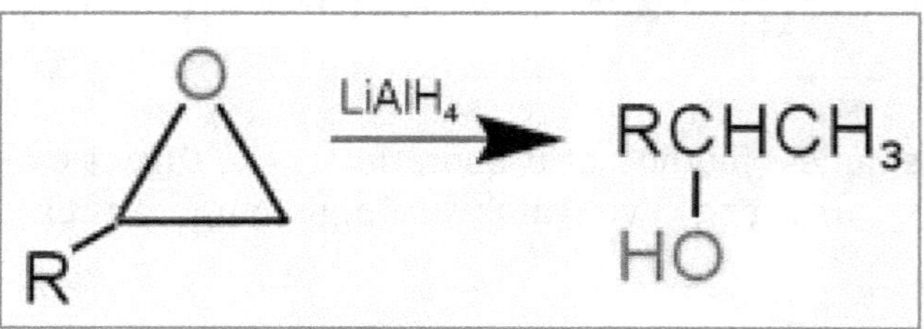

e) Oxiranes form **trimeric component** (polymeric ether) 1,4,7 - trioxycyclononane under the effect of antimony pentafluoride.

f) Oxirane releases the ring strain by leading to the opening of the ring by many reagents and forming respective alcohols.

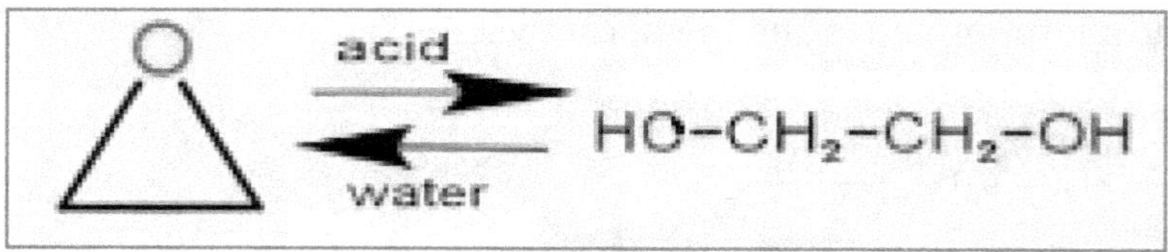

g) Oxirane substituted with cyanide ions show ring opening.



h) **Friedel - Crafts Reaction** leads to the formation of phenethyl alcohol.

i) Synthesis of crown ethers

$$n(CH_2CH_2)O \longrightarrow -(CH_2\text{-}CH_2\text{-}O)_n$$

$$SO_2 + 3 \triangle \xrightarrow{Cs^+}$$

**11-membered hetero-cyclic compound**

## APPLICATIONS OF OXIRANE

a)      **Leukotrienes** as implied in branchial asthma and **Disparlure** is used by insect for sex attraction.



b)      **Methyl oxirane** finds usage in detergents and thermoplastics. Oxirane, many a times, show natural occurrences having anticancer and antibiotic properties.

## 5.2.3 THIIRANES

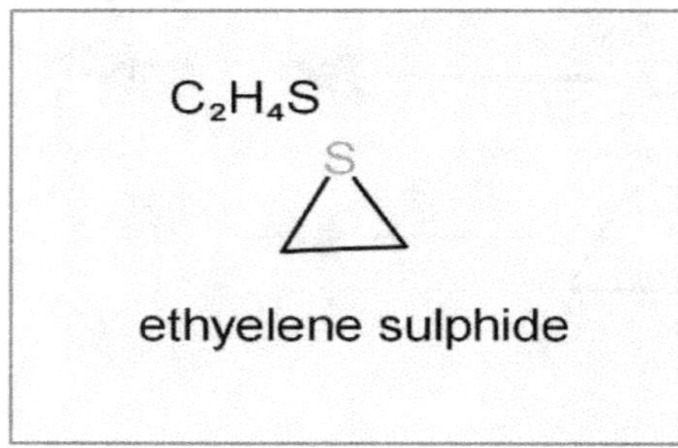



## SYNTHESIS OF THIIRANE

a)      Thiiranes are popularly prepared from 2-mercaptoethanol in the presence of phosgene (COCl$_2$) and pyridine.

b)      Thiiranes can readily be obtained from oxiranes by reaction with thiocyanate ions. Cyclic intermediates are presumably formed in the reaction.



## Mechanism

**Step 1-** Thiocyanate ion attacks the three - membered heterocyclic ring and results in ring opening.

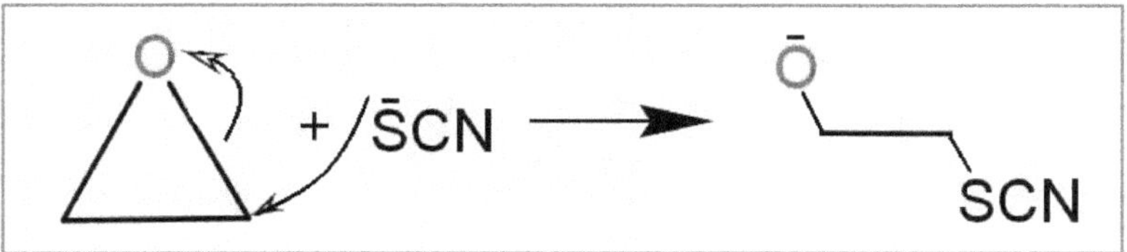

**Step 2**-In the ring formation step, a cyclic intermediate is formed.

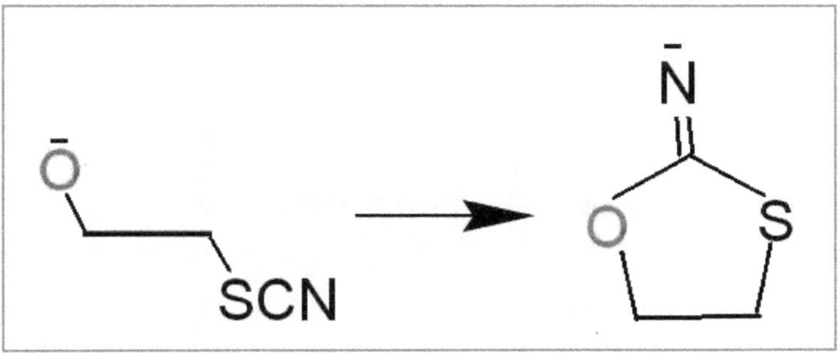

**Step 3**- The cyclic intermediate cleaves.

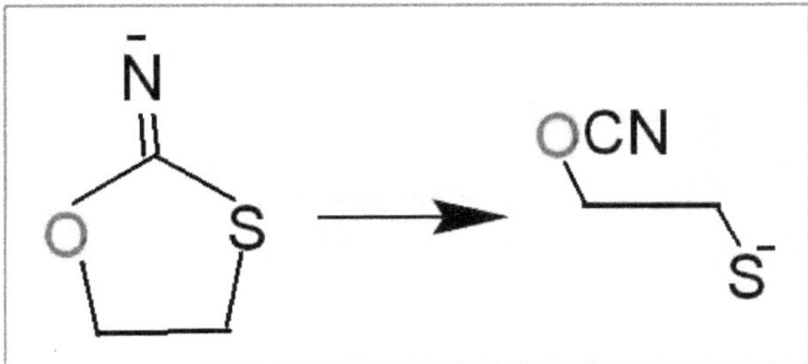

**Step 4**- After two displacements and **inversion** in configuration, the desired product is obtained.

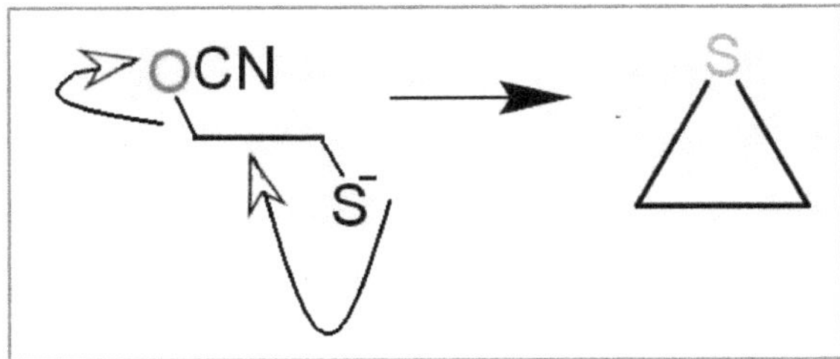

c)      KSCN and ethylene carbonate react to form thiirane.

$$KSCN + C_2H_4O_2CO \longrightarrow KOCN + C_2H_4S + CO_2$$
$$\text{thiirane}$$

## PROPERTIES OF THIIRANES

a)      Thiirane is a pale-yellow liquid with boiling point 56°C; melting point -109°C.

b)      Thiirane ring is **cleaved** by good nucleophiles. Weak nucleophiles also yield the same ring opening products if accompanied by acid catalysts.

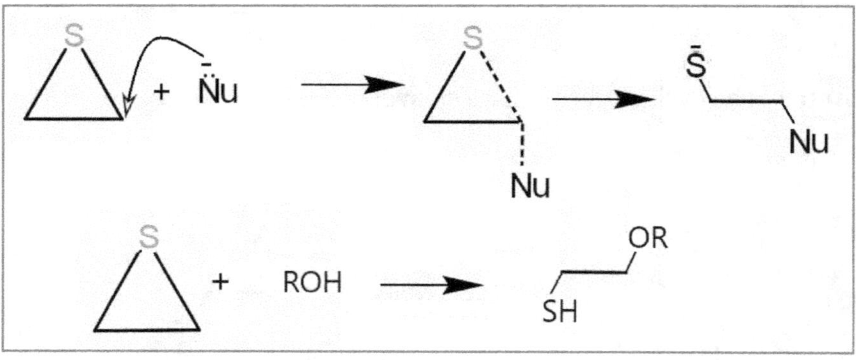

c)      Chloro-mercaptan is formed as a result of **ring opening** when benzo-fused thiirane reacts with HCl.



d)      Thiirane shows removal of Sulphur (**Desulphurization**) under various conditions:

I. When thiirane reacts with methyl iodide, sulphur atom is removed to finally form an alkene via epi-sulphonium salt intermediate.

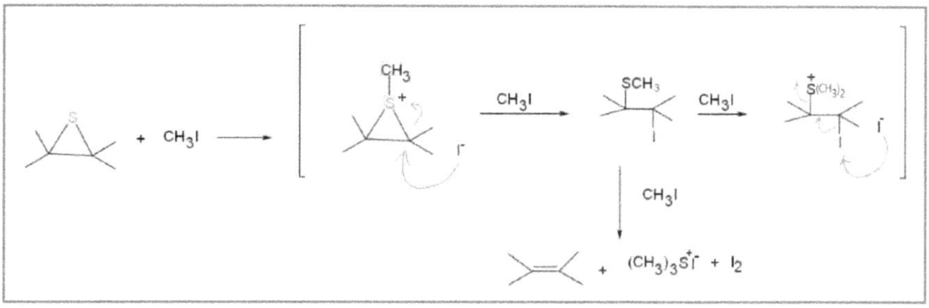

II. If the substituents on the carbons of thiiranes are aryl, then removal of Sulphur can be brought about by heat and alkene is obtained.

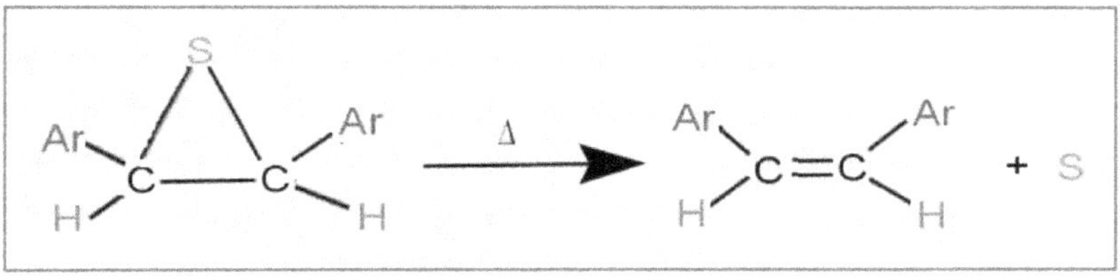

e) Treatment of thiirane with ultraviolet light in the presence of **fumaronitrile** results in formation of alkene after removal of Sulphur.

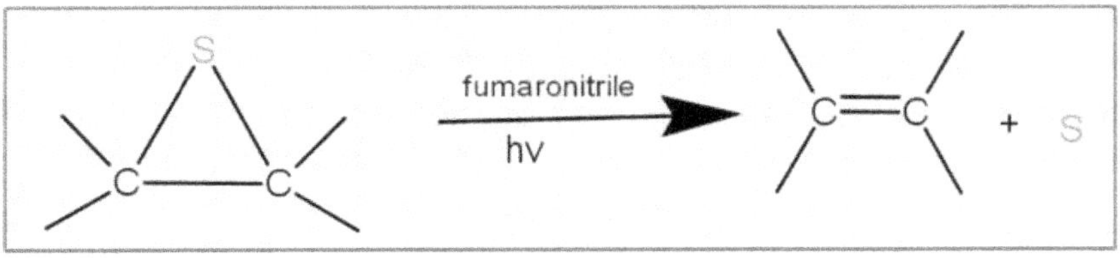

However, if tetracyanoethylene $(C_2(CN)_4)$ is used in place of fumaronitrile then **tetrahydrothiophene** is obtained in presence of ultraviolet light.

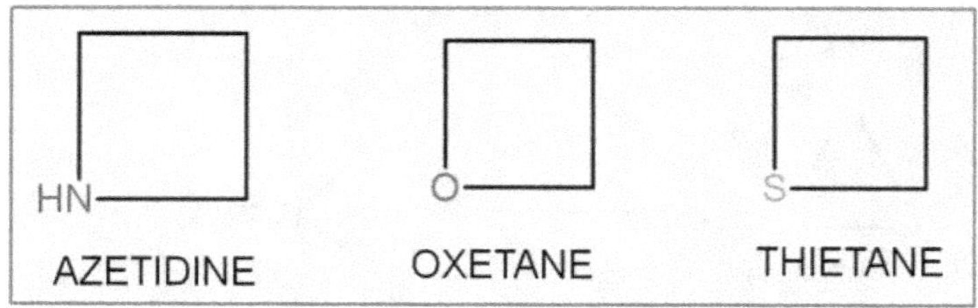



## APPLICATIONS OF THIIRANE

It is the simplest episulphide and the simplest sulphur containing compound. Thiirane is used as an organo-sulphur compound. It is also used to illustrate the main ethylene sulphide derivatives.

## 5.3 4- MEMBERED HETEROCYCLIC COMPOUNDS



## GENERAL METHODS OF PREPARATION

Certain methods of synthesis of four-membered heterocycles have been proposed; low yield products are obtained on treatment of 3-halo alcohols, amines, or thiols with base; however, dimers and elimination products cannot be avoided. If weak base is used, then thietane derivatives can easily be obtained due to greater nucleophilicity of sulphur.



If strong base is used and oxygen and sulphur atoms are eliminated in substituents, then only Azetidine is obtained.

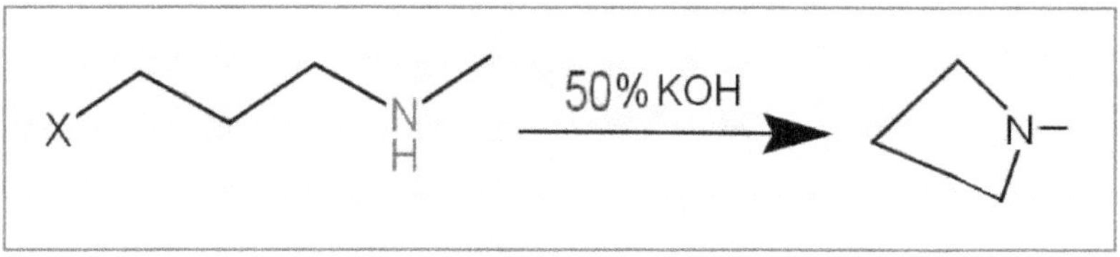

Oxetanes can be formed by **Paternò–Büchi photocyclization** in which a carbonyl compound and an alkene react together in the presence of photon energy. It is a photocycloaddition reaction with carbonyl compound in excited state and alkene in ground state.

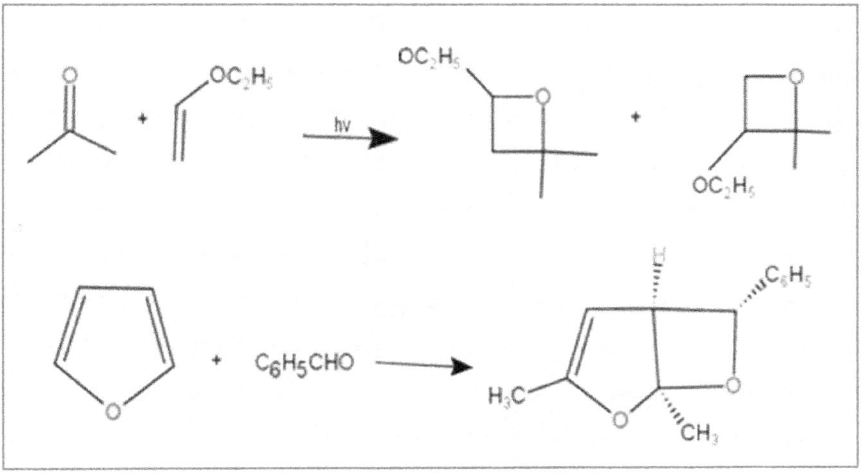

## GENERAL REACTIONS

Ring Strain influences majority of reactions of four membered heterocycles. Ring cleavage can be obtained if acid is used as a catalyst.

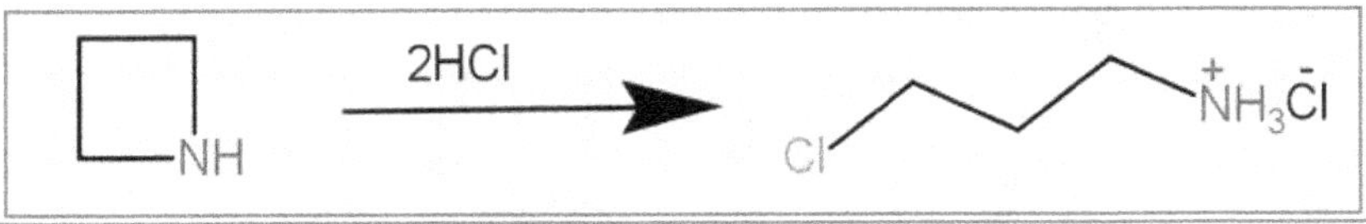

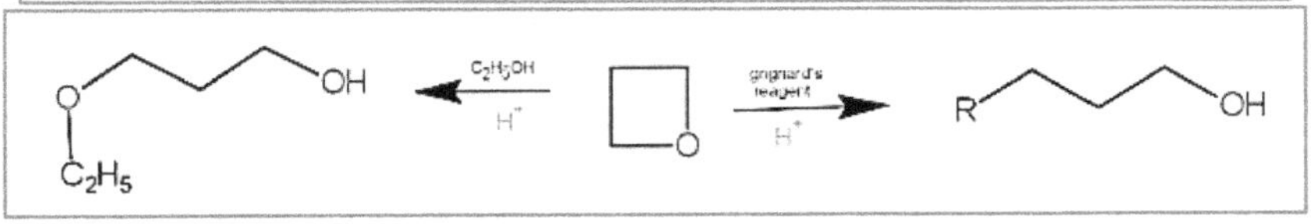

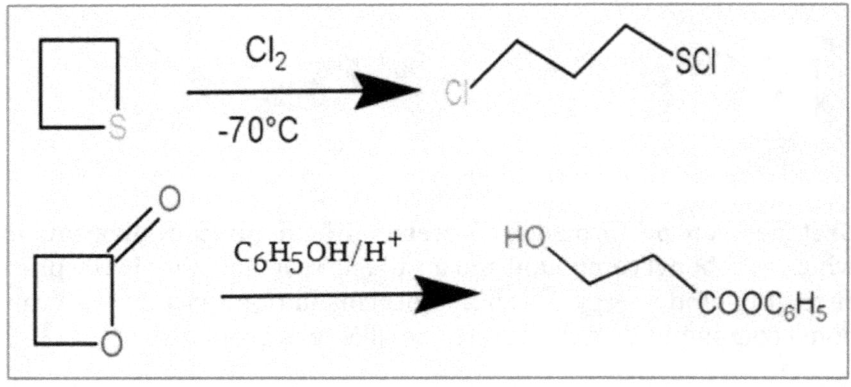

## 5.3.1 AZETIDINE - $C_3H_7N$

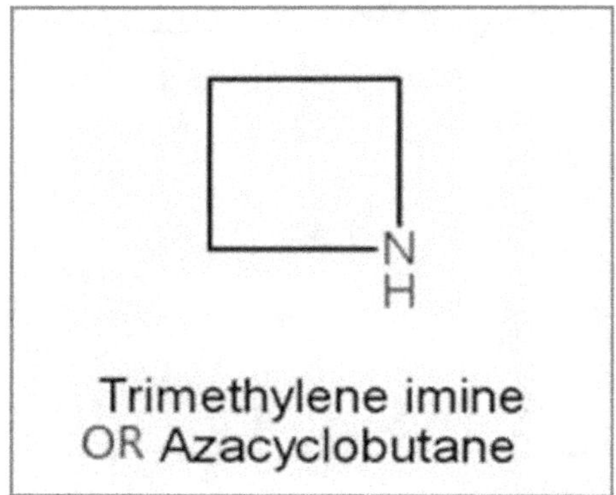



**SYNTHESIS**

a)      Phosphine imide (A) on **cyclization** results in Azetidine formation where thermal elimination of triphenylphosphine oxide cannot be avoided.

b) **Cycloaddition** of tri-methylene chlorobromide and p-toluene Sulphonamide results in Azetidine formation through a series of reactions.

In **Step 1**, the two reactants, in the presence of an alcohol, result in cyclic adduct formation.

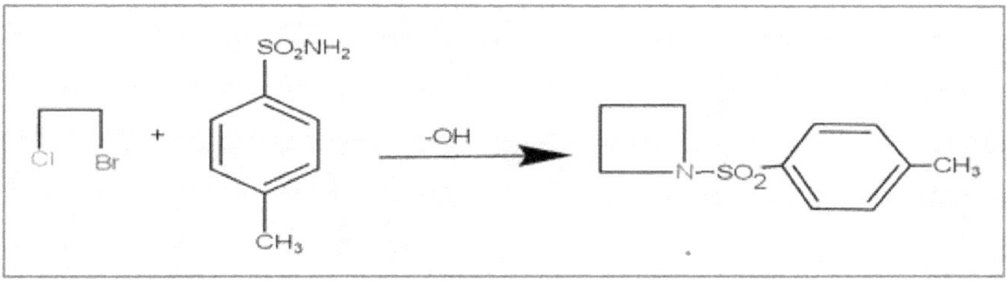

**Step 2** involves reduction in the presence of sodium and n-pentanol to yield the desired product.

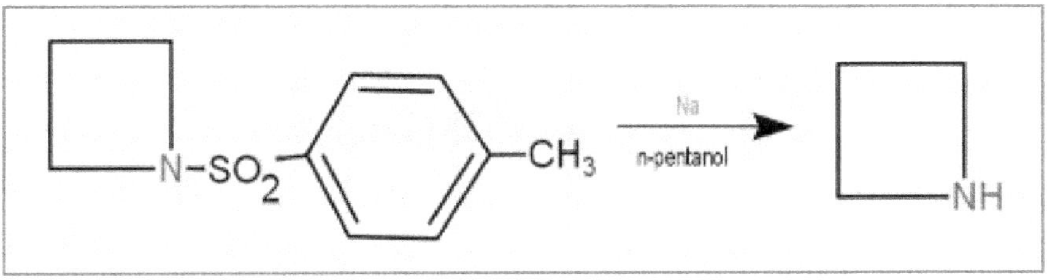

c) Similarly, if sodium and n-pentanol couple is used with tosyl chloride and pyridine on iso-oxazoline then Azetidines can be obtained.

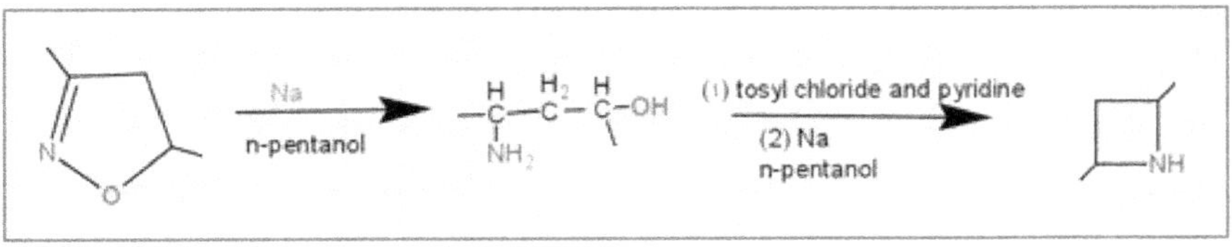

## PROPERTIES

a) Also known as Azacyclobutane, Trimethylene imine and 1,3 - propylenimine.
b) Molecular weight - 57.09 gm/mol

68

c) Clear pale brown liquid.

d) Miscible with water and has a density of 0.847gm/cm³ at 25°C.

e) Boiling point = 61 – 62°C

f) Action of **nucleophiles** and bases show negligible response by Azetidine but when it is exposed to hydrogen peroxide it results in the formation of acrolein and ammonia derivative showing opening of ring.

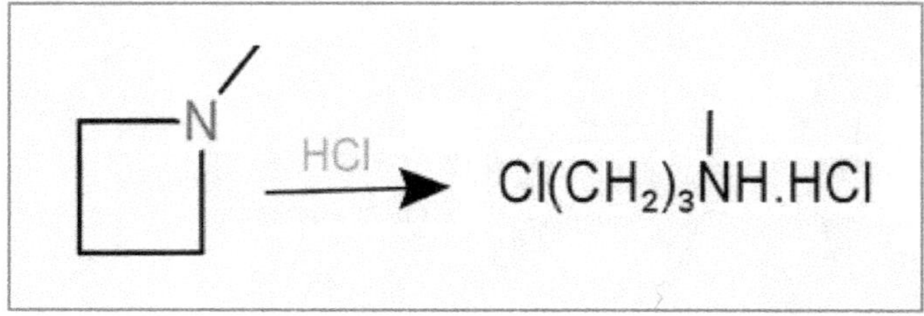


g) **Electrophiles** show ring cleavage at high rate of reaction.

h) Some addition reactions are also exhibited by Azetidine with carbon disulphide, salt is formed; Formaldehyde yields N-hydroxymethylazetidine and a quaternary salt is obtained with alkyl halide.

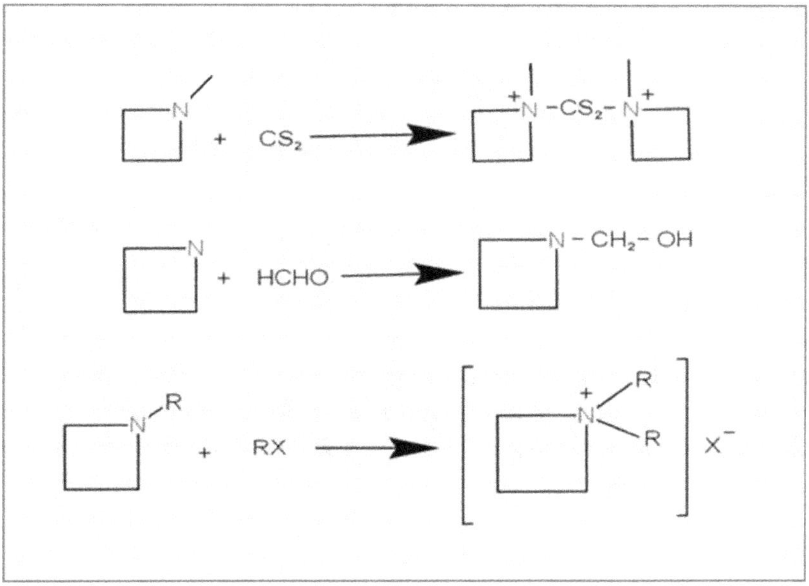

## APPLICATIONS OF AZETIDINES

Azetidine is a four membered nitrogen containing heterocycle saturated in nature used as building block in drug design on account of its reasonable chemical stability. **Azelnidipine** is used as an antihypertensive agent. It is a dihydropyridine calcium channel blocker which produces a long-lasting decrease in blood pressure.

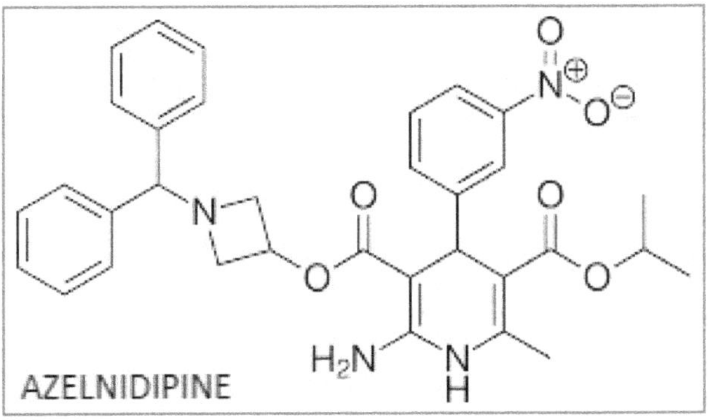

AZELNIDIPINE

### 5.3.2 OXETANE - $C_3H_6O$

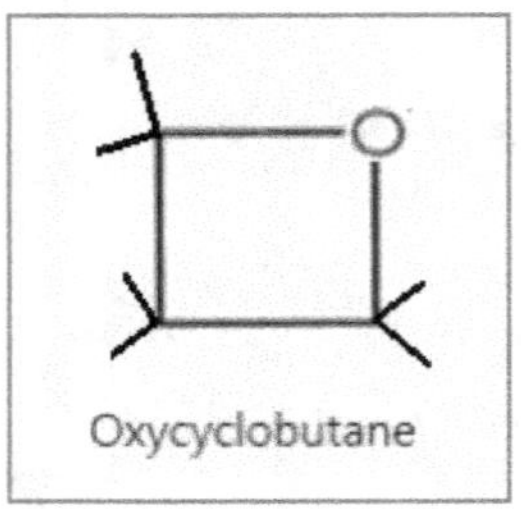



Oxetanes are four-membered heterocycles with oxygen as heteroatom where the ring is planar but due to difference in carbon-carbon (1.54 Å) and carbon-oxygen (1.46 Å) bond length, oxetanes are not perfect squares.

### SYNTHESIS

a)      1,3-Halohydrins slowly cyclizes **intramolecularly** in the presence of base to yield oxetanes. However, prior to base treatment if acetylation is performed, then rate of reaction increases.

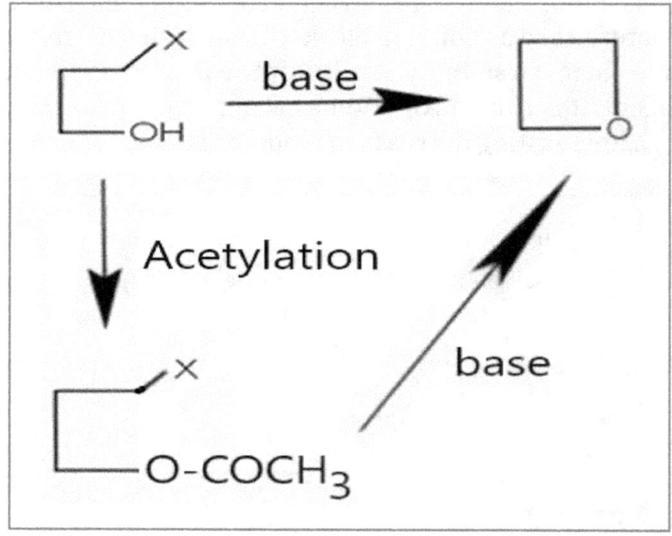



b)      Carbonyl compounds and alkenes undergo addition reaction in the presence of light (hν energy) via **Paternò–Büchi** reaction to result in the formation of oxetanes.

The reaction mechanism proceeds via stable **free- radical interme- diate** formation.

## PROPERTIES

a) Synonyms of oxetane are 1,3- Epoxypropane, Oxacyclobutane, 1,3-propylene oxide and tri-methylene oxide.
b) Molecular weight = 58.08 gm/mol.
c) Melting point = - 97°C
d) Boiling point = 49 - 50 °C.
e) Density = 0.8930 gm/mol.
f) **Ring cleavage** of oxetane can be easily brought about in the presence of acid. 2-methyloxetane undergoes ring opening in the presence of hydrochloric acid to give substituted carbon chain products.

g) Aryl or alkyl substituted oxetanes undergo **Friedel- Crafts reaction** to yield ring opening products.

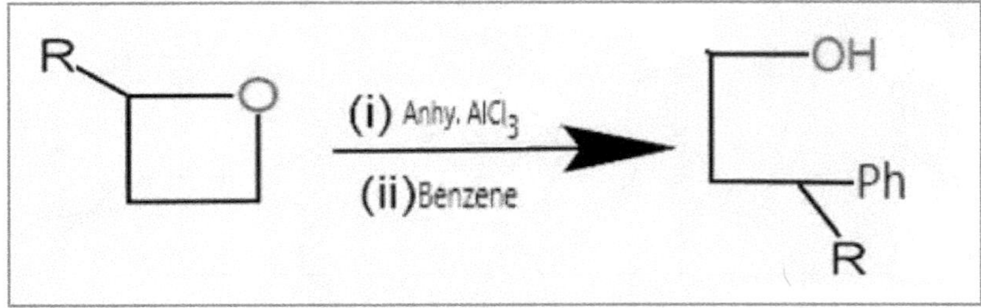

h) Carbonates are formed when oxetanes respond to carbon - di- oxide in the presence of tetra-phenyl antimony iodide.

trimethylene
carbonate

i) In the presence of Grignard's reagent, the oxetane opens up the ring and forms straight chain alcohols.

## APPLICATIONS OF OXETANE

**Paclitaxel** is a chemotherapy medication used to treat several types of cancer. These includes lung cancer, breast cancer, cervical cancer, and pancreatic cancer.

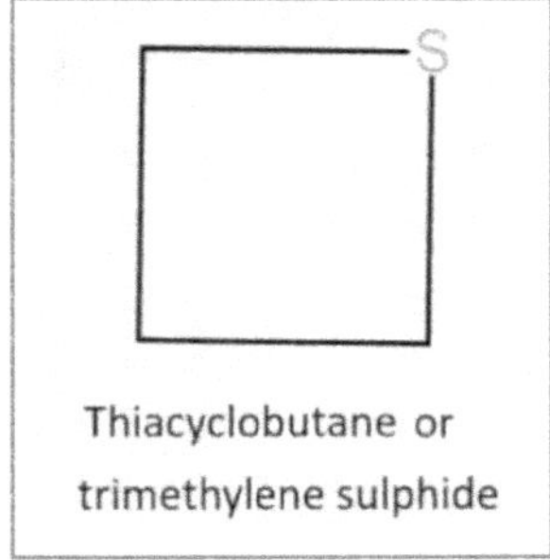


### 5.3.3 THIETANE - $C_3H_6S$

Four- membered heterocyclic compounds that have Sulphur as heteroatom and three -CH$_2$ groups are called Thietane.

**SYNTHESIS**

(a)    **Cyclisation** of dihaloalkanes with Sodium sulphide (anhydrous) in the presence of ethanol at high temperature results in the formation of Thietane.

$$XCH_2 - CH_2 - CH_2X \;+\; Na_2S \xrightarrow[\text{heat}]{\text{EtOH}}$$

74

(b)  When substituted or unsubstituted 1-bromo-3-chloropropane reacts with thiourea under alkaline conditions, substituted or unsubstituted Thietane can be obtained.

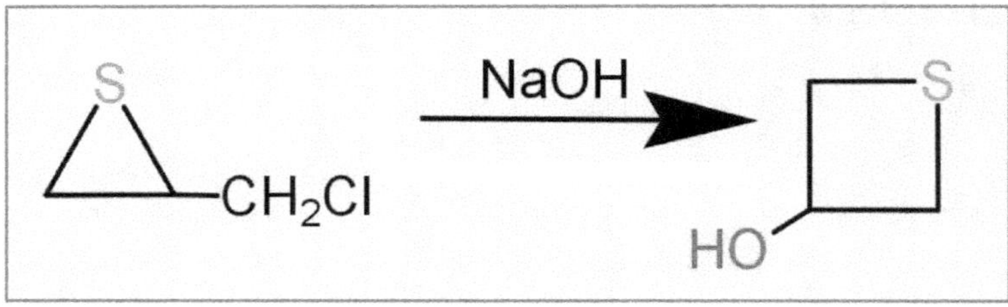

(c)  Alkaline conditions first bring about **ring opening** in chloromethyl thiirane and then cyclization to from 3- hydro thietane.

**PROPERTIES**

a) Thietane is also known as Thiacyclobutane and trimethylene sulphide.
b) It is a colourless liquid with a sulphurous odour.
c) Molar Mass = 74.14 gm/mol.
d) Boiling point = 94 - 95°C.
e) Density = 1.028 gm/cm$^3$.
f) As compared to the three-membered compound containing Sulphur as hetero atom (thiirane) the four membered compound (thietane) witnesses a **decrease in ring strain** and as a result show breaking of ring either slowly or under drastic conditions.

g) 3-Aminopropanethiol is obtained on treatment of Thietane with ammonia in a sealed tube at 200°C.

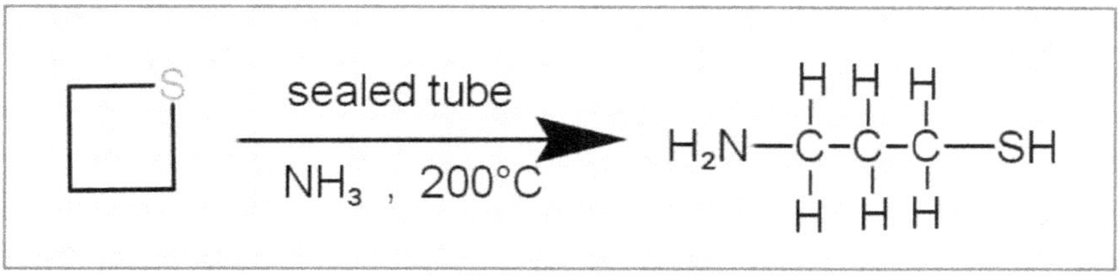

$$H_2N-CH_2-CH_2-CH_2-SH$$

h) **Oxidation** of Thietane occurred easily in the presence of $H_2O_2$ to yield sulphone.

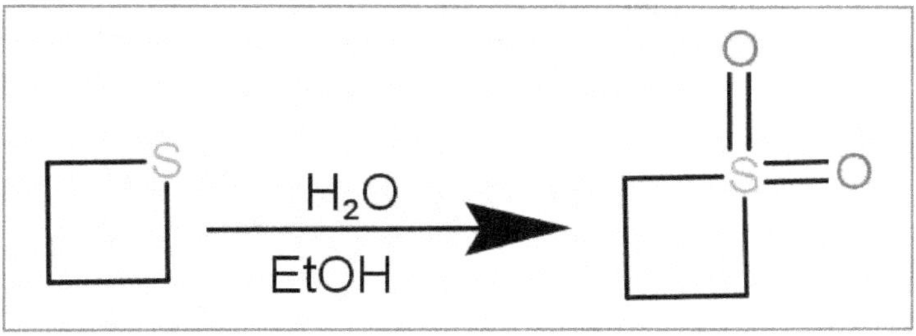

i) Geometrical i.e., **cis- trans isomerism** in exhibited by Thietane on formation of alkene during **photochemical reaction**.

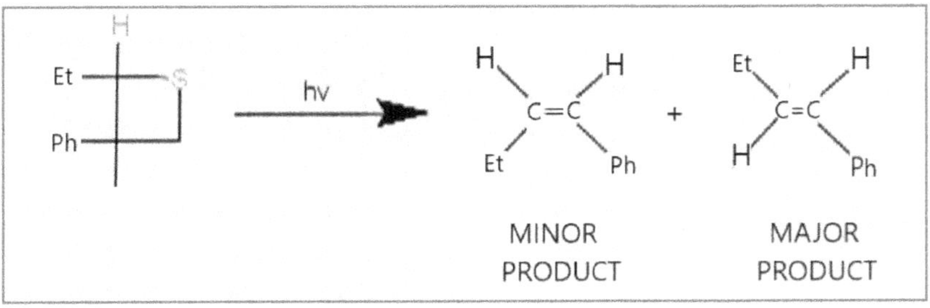

MINOR PRODUCT

MAJOR PRODUCT

j) An intermediate, quaternary salt in obtained when Thietane is treated with methyl iodine which soon opens to a straight chain ionic compound.

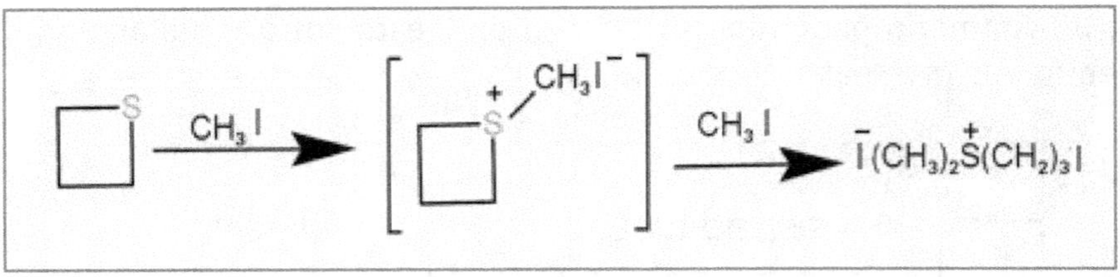

j) Thietane-3-one, on reduction with NaBH$_4$, gives 3-thietanol which opens up in the presence of Raney Nickel to yield isopropyl alcohol.

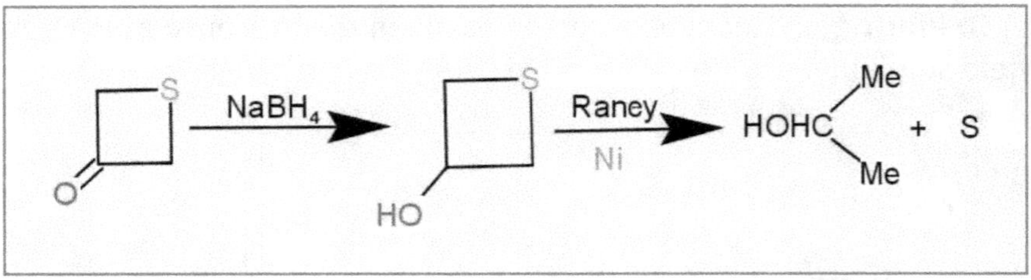

## APPLICATIONS OF THIETANE

- Mouse alarm pheromones and predator scent analogues are major applications of thietanes and its derivative 2 - propyl thietane.

- **Alitame** is a high intensity, sweetener formed from the amino acids, l-asparatic acid and d- alanine, and an amine derived from thietane. It has an excellent stability at high temperature and can be used in cooking and baking.

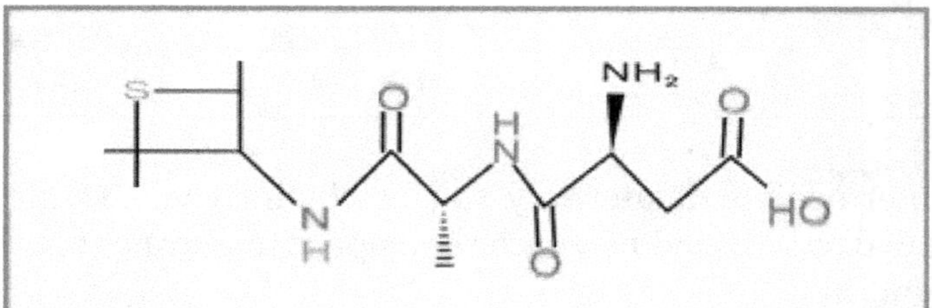

Alitame

# REFERENCES


Tanner, D. (1994) Angew chem, Int. Ed. Engl, 33, 599.

Bartok, M. and Lang, K.L. (1985), Small ring Heterocycles, Part **3**, ed. A. Hassner, Wiley Interscience, New York, p.1.

Jorgensen, K. A. (1989), chem. Rev., 89, 431.

Zoller, U. (1983), Small Ring Heterocycles, Part **1**, ed. A. Hassner, Wiley Interscience, New York, 1983, p.333.

Searles, S., Lutz, E., Hays, H.R. and Mortensen, M.E. (1962), Ethylene Sulfide, 42, 59.

Van Tamelen, B. E. (1951), J. Am. Chem. Soc., 73, 3444.

Helm Kamp, G.K. and Pettitt, D. J. (1960), J.Org. Chem., 25, 1754.

Culvenor, C. C. J., Davies, W. and Heath, N. S. (1949), J. Chem. Soc., 282.

Kamata, M. and Miyashi, T. (1989), J. Chem. Soc., Chem. Commun., 557.

Davies, D. E. and Storr, R. C. in A. R. Katritzky and C. W. Rees (1984) (Eds.) Comprehensive heterocyclic chemistry Vol. **7**, Pergamon press, Oxford, 238.

Moore, J. A. and Ayers, R. S. in A. Hassner (1983) (Ed.) Small Ring Heterocycles, Part 2nd Wiley - Interscience, New York, 1.

Searles, S., Tamres, M., Block, F. and Quarterman, L. A. (1956), J. Am. Chem. Soc., 78, 4917.

Schaefer, F. C. (1955), J. Am. Chem. Soc., 77, 5928.

Freeman, J. P., Pucci, D. G. and Birsch, G. (1972), J. Org. Chem., 37, 1894.

Searles, S. and Butler, C. F. (1954), J. Am. Chem. Soc., 76, 56.

Gupta, R. R., Kumar, M. and Gupta, V. (2013), Heterocyclic Chemistry Vol.**1**, Springer-Verlag Berlin Heidelberg, 364.

Ditters, G. and Muellar, E. (1965) (Ed.), Methoden der Organischen chemie, Houben-Weyl, 493; Searles in A.R. Katritzky and C.W. Rees (1984) (Eds.) Comprehensive Heterocyclic Chemistry Vol.7, Pergamon Press, Oxford, 263.

Seales, Jr., S., Pollart, K.A. and Block, F. (1957), J. Am. Chem. Soc. 79, 952.

Dittmer, D. C., Hertler, W. R. and Winicov, H. (1957), J. Am. Chem. Soc., 79, 4431.

Pasterno, E. and Chieffi, G. (1909), Gazz. Chem. Ital., 39,342.

Arnold, D. R. (1968), Adv. Photochem., 6, 301.

Gotthardt, H., Steinmetz, R. and Hammond, G. S. (1968), J. Org. Chem., 33, 2774.

Arnold, D. R. and Glick, A. H. (1966), Chem. Commun., 813.

Matty, S. and Buchkremer, K. (1985), Heterocycles 27, 2153, 1988; M.Braun, Nachr. Chem. Techn. Lab., 33, 213.

Arnold, D. R., Hinman, R. L. and Glick, A. H. (1964), Tetrahedron Lett. 1425.

Scarles Jr., S. in A. Weissberger (1964) (Ed.), The Chemistry of Heterocyclic Compounds 19, Part 2nd, Wiley- interscience, New York, 983.

Yamaguchi, M., Nabayashi, Y. and Hirao, I. (1984) Tetrahedron 40, 4261.

Suzuki, T., Saimoto, H., Tomioka, H., Oshima, K. and Nozaki, H. (1982), Tetrahedron Letters, 3597.

Hudrlik, P. F. and Wan, C. N. (1975), J. Org. Chem., 40, 2963.

Gupta, R. R., Kumar M. and Gupta, V. (2013), Heterocyclic Chemistry, Vol. **1**, Springer- Verlag Berlin Heidelberg, 389.

Sander, M. (1984), Chem. Rev. 66, 341, 1966; A.R. Katritzky and C.W. Rees (Eds.) Comprehensive Heterocyclic Chemistry, Vol. 7, Pergamon press, Oxford, 403.

Dittmer, D. C., Segerman, T. C. and Hassner, A. (1985) (Ed.) Small Heterocycles part 3rd, Wiley-interscience, 431.

Sander, M. (1966) Chem. Rev. 66, 341; E. Block of A.R. Katritzky and E.W Rees (Eds.) Comprehensive Heterocyclic Chemistry, Vol. **7** Pergamon press, oxford, 403.

Baker, H. J. and Keunig, K. J. (1934); Rec., Trav. Chem., 53, 808.

Etienne, Y., Soulas, R. and Lumbroso, H. in A Weissberger (1964) (Ed.), The Chemistry of Heterocyclic Compounds XIX, Part 2nd, Wiley - Interscience, New York, 647.

Bost, R. W. and Conn, M.W. (1933), Oil Gas J., 32, 17.

Bordwell, F. G. and Pitt, B. M. (1955), J. Am. Chem. Soc., 77, 572.

Adams, E. P., Ayad, K. N., Doyle, F.L, Holland, D. O., Hunter, W.M., Nayler, J. H. C and Queen, A. (1960), J. Chem. Soc., 2665.

Gupta, R. R., Kumar, M. and Gupta, V. (2013), Heterocyclic Chemistry, Vol.1, Springer- Verlag Berlin Heidelberg, 402.

Rice, D. R. and Stier, R. D. (1973), Chem. Comm., 166.

Palmer, D. C. and Taylor, E. C. (1986), J. Org. Chem., 51 ,846.

Trost, B. M., Schinski, W. L. and Mantz, I. B. (1962), J. Am. Chem. Soc., **91**, 4320.

Thorbjorn, S. and Matthias, L. (2016), Chem. Senses., **5**, 41, 399-406.

Brechbuhl, J., Klaey, F., Nenninger-Tosato, M., Humi, M., Sporkert, N., Giroud, F., Broillet, C. (2013), Proc. Natl. Acad. Sc., USA 110 (**12**), 4762-4767.

# CHAPTER 6

## 5-MEMBERED HETEROCYLIC COMPOUNDS

### 6.1 INTRODUCTION



The parent heterocyclic five membered compound (1) is pyrrole (X=NH), furan (X=0) and thiophene (X=S) accordingly. The benzo-fused heterocycles can be regarded as heterocyclic analogues of naphthalene and dibenzo heterocycles (4) bear similarity to phenanthrene.

### 6.2 GENERAL METHOD OF SYNTHESIS

Ring synthesis in which a bond is formed between the hetero atom and a carbon atom are considered according to whether the heteroatom functions as a nucleophile, an electrophile, a radical or other electron deficient species. The simplest of ring closure depending on nucleophilic property is based on the **intramolecular displacement** of a suitable leaving group.



There are a variety of examples of ring formation in which the oxygen atom of a carbonyl group functions as a nucleophile.

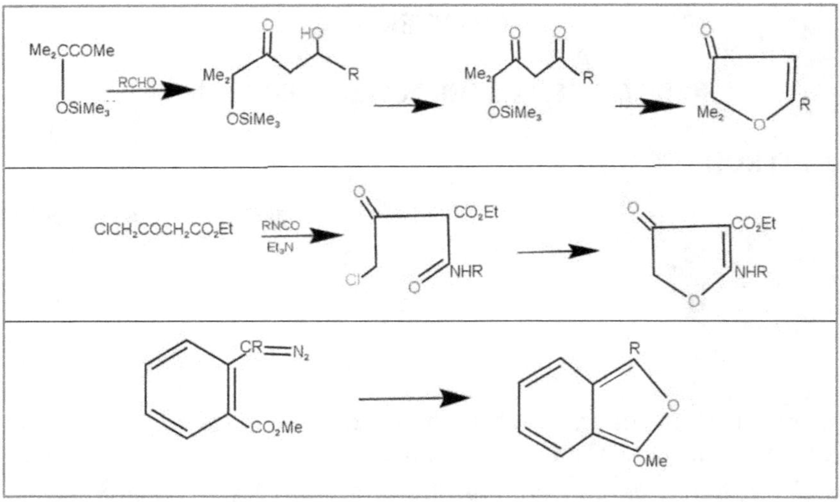

The **intramolecular addition** of the heteroatom to a suitably disposed double bond is the basis of a growing variety of the ring synthesis. The **cyclization** of O- allylphenols to 2,3-dihydrobenzofurans(a) accompanies the Claisen rearrangement of allyl aryl ethers and is best promoted on a preparative basis by acid catalysis. 2,3- dihydro-benzo thiophene and 2,3-dihydrobenzoselenophenes have been obtained through analogous rearrangements. Cyclization of O-(2chlorophyallyl) phenols leads directly to 2-methyl benzofurans (b).

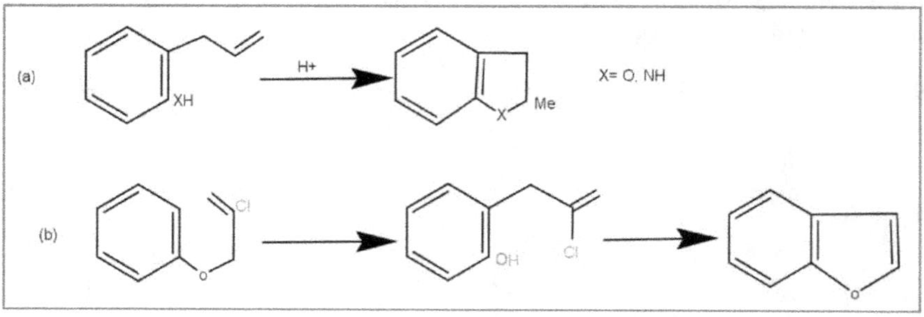

The treatment of O-methyl seleno cinnamates with bromine in pyridine gives excellent yields of benzoselenophene-2-carboxylates. There are several other **ring closure** reactions that suffice as good examples for synthesis by ring closure.

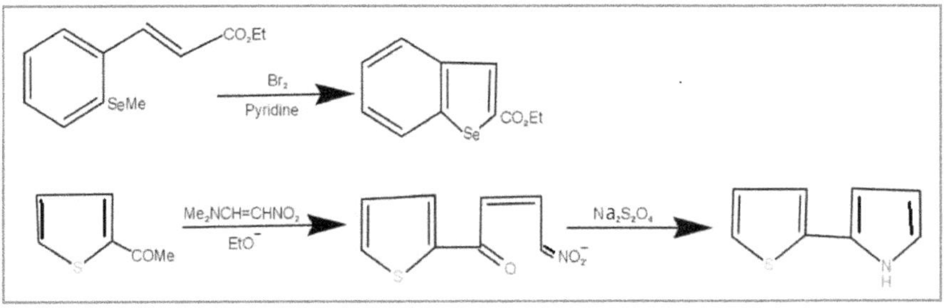

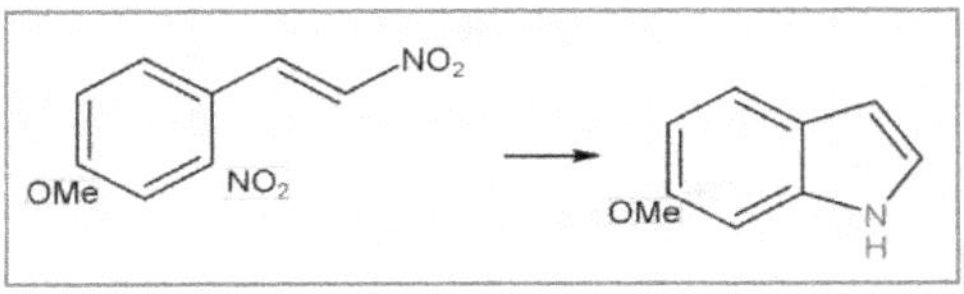

[in presence of $H_2$/Pd-C]

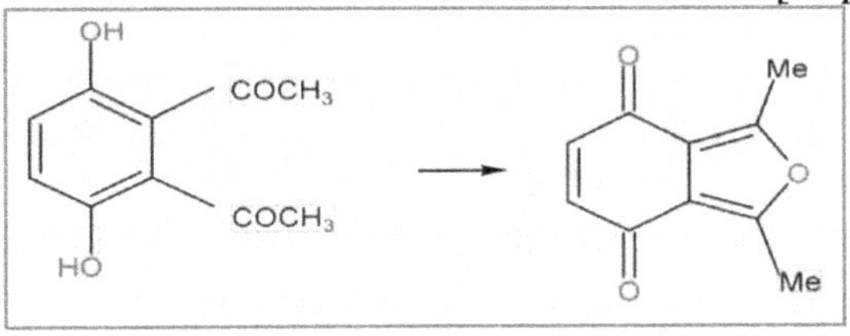

[ in presence of $80\%H_2SO_4$]

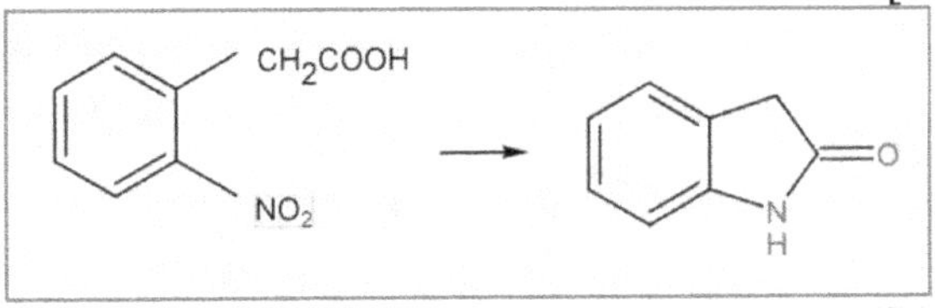

[in presence of $H_2$/Pd-C]

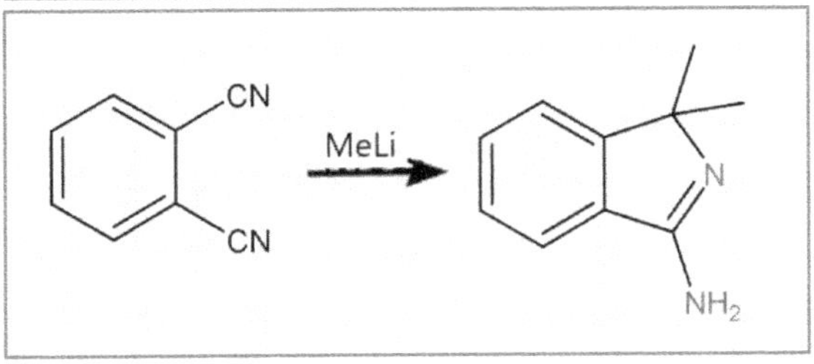

**Intramolecular carbonian addition** to a nitrile group provides access to 3-amino heterocycles. Similarly, intramolecular addition of the carbanion to an ester group provides 3-hydroxy heterocycles.

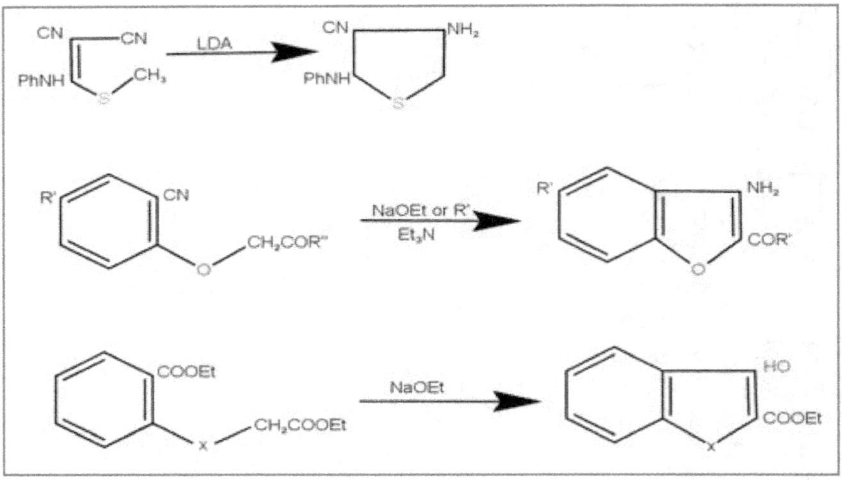

82

Cyclization initiated **photochemically** leads to the conversion of diphenyl ethers to dibenzofurans diphenyl amine to carbazole. The oxidative ring closure of diphenyl amine to carbazole is also affected by palladium (II)acetate.

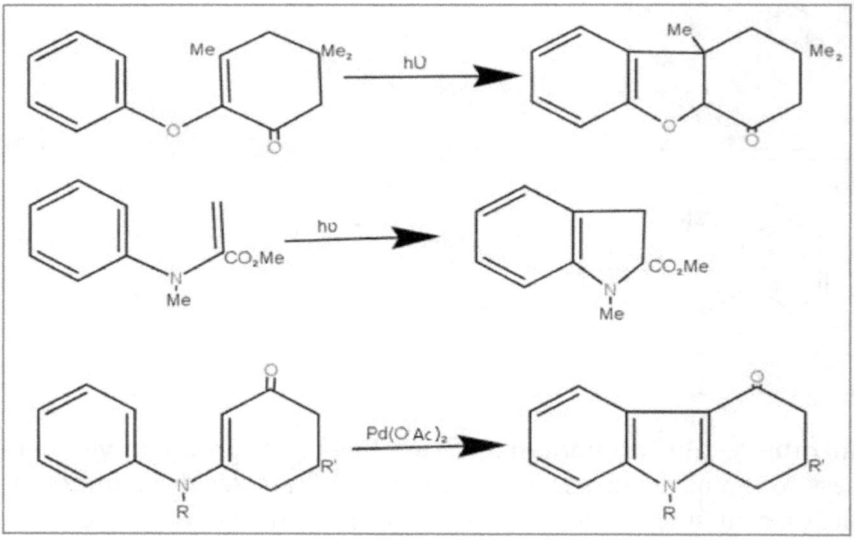

Ring formation by **intramolecular acylation** is shown by the isatin synthesis. In the same way formal 1,4-methyl group migrations have been observed in the cyclization of mesityl hydrazones.

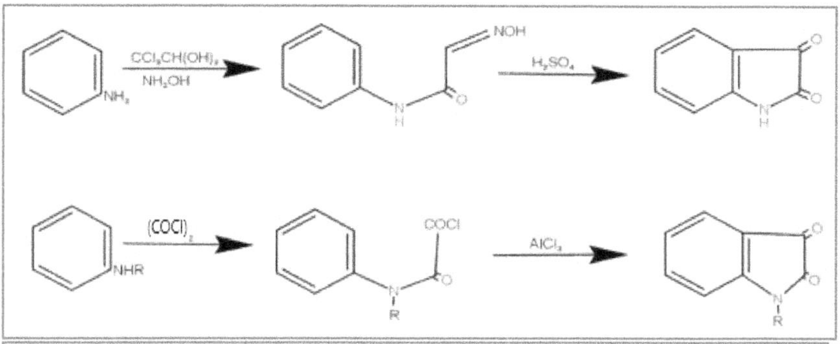

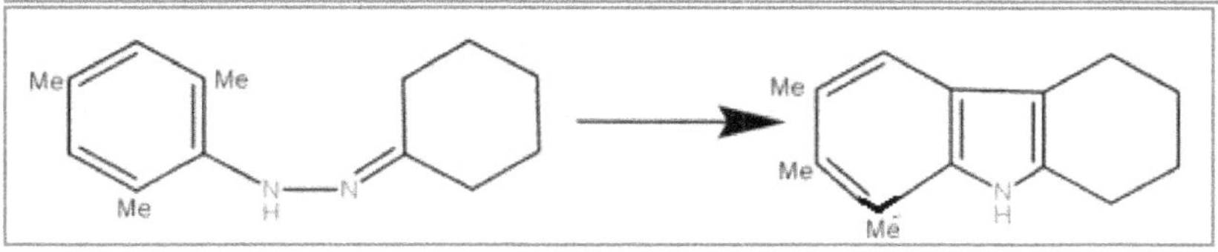

In yet other cases, the ring formation either requires an **internal oxidation(a)** or the use of a reducing agent (**b, c**). Addition reaction with tetracyano ethane have provided access to 2,5-diamino-3,4-dicyano-thiophene and selenophene. Base catalysed **rearrangement** gives the isomeric pyrrole thiol (**d**).

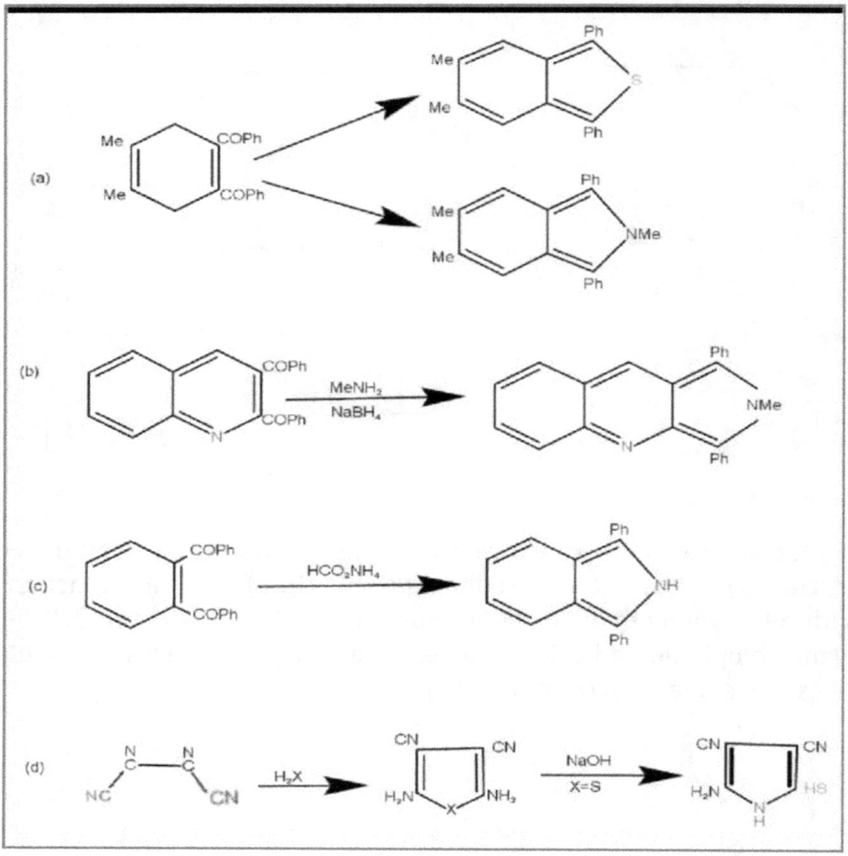

(a)

(b)

(c)

(d)

There are certain **ring contraction** reactions that proceed through intramolecular displacement.

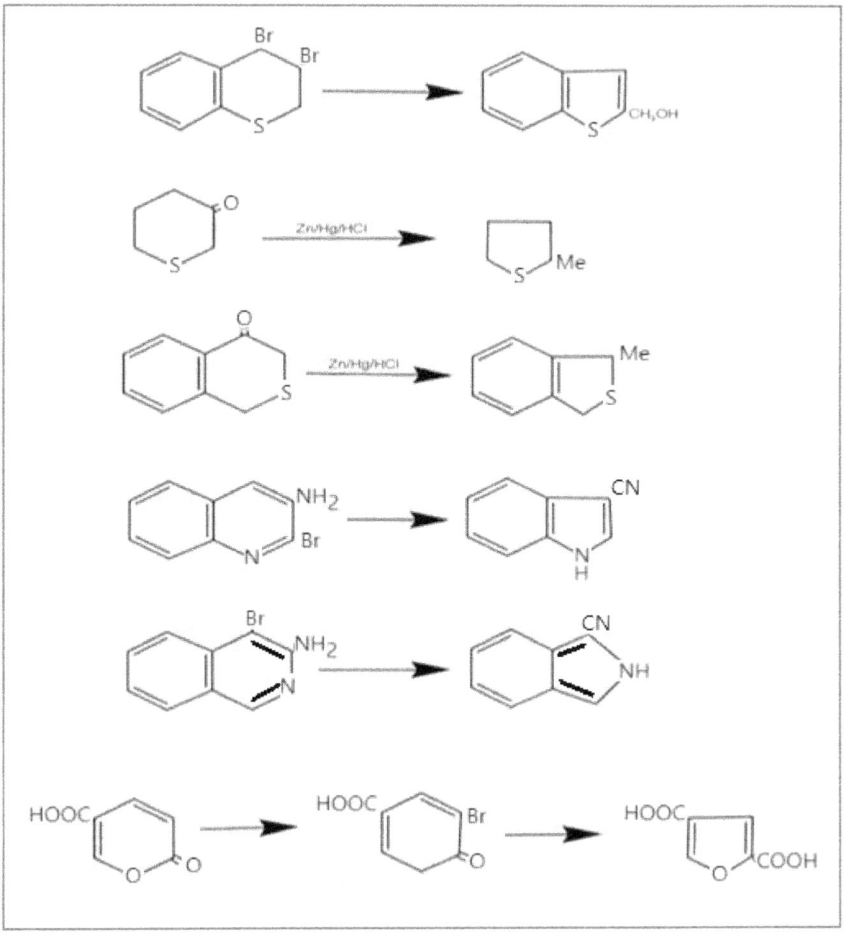

## 6.3 GENERAL CHEMICAL PROPERTIES

The promoted heterocycles are highly reactive electrophilic species which by reaction with the conjugated base give dimers, trimers and even polymers. Pyrroles are especially sensitive to oxidation and N-unsubstituted derivatives are readily converted to reactive anions by abstraction of proton bound to nitrogen.

a) <u>Reactions involving heat and light:</u>

The conversion (Rearrangement) of N-substituted pyrroles to C-substituted is the resultant of thermal reaction. **Flash vacuum pyrolysis**

86

(FVP) of 2-methoxycarbonyl pyrrole gives the ketone; on warming the dimer is formed.

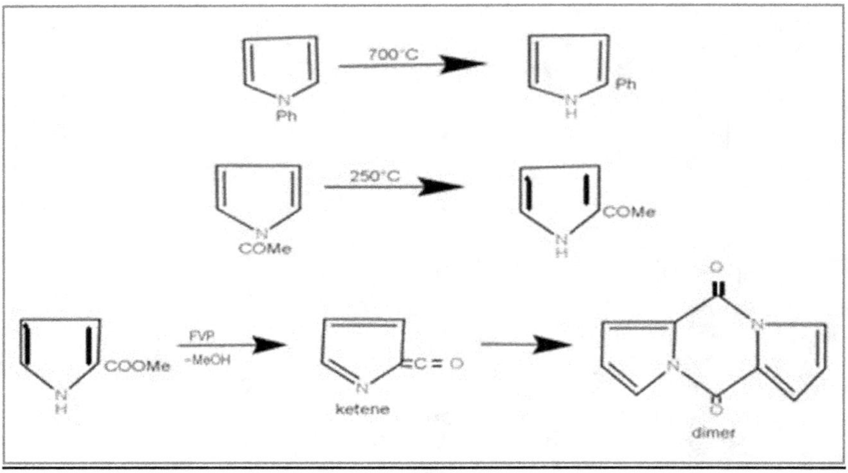

**Photodimerization** reaction is also of mention in indoles and iso-benzo furan.

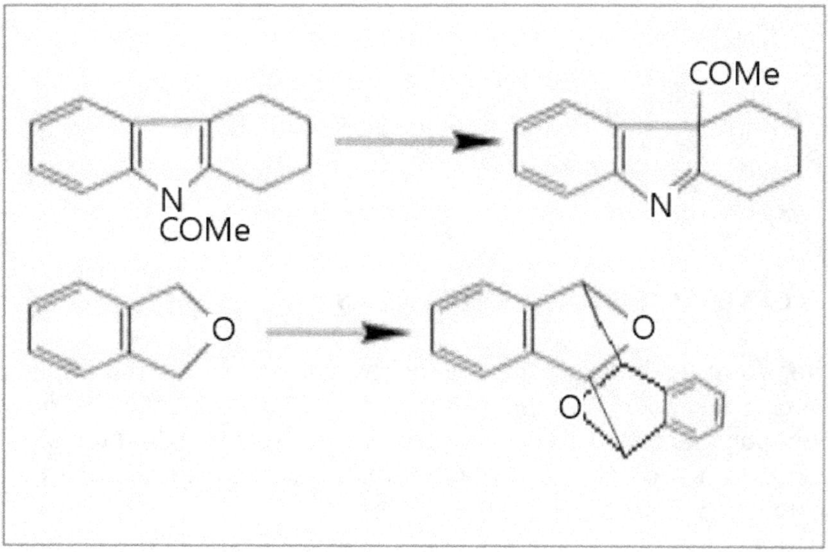

b) <u>Reaction involving electrophiles:</u>

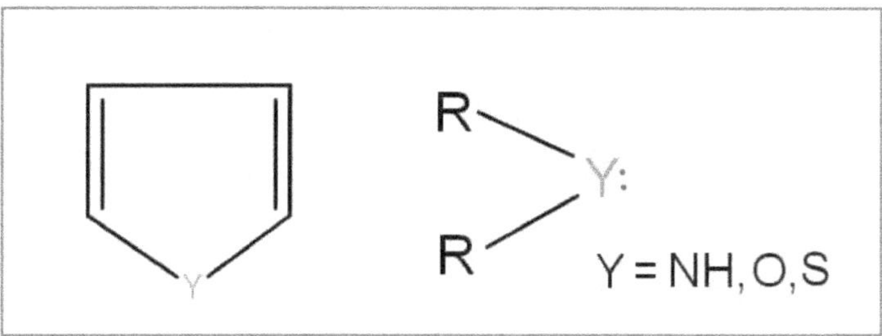

The structure of pyrrole, furan and thiophene indicates that these compounds may show chemical properties like those of amines, ethers and thioethers. The starting attack of the electrophile would be expected to take place at heteroatom and thereby form quaternary ammonium and oxonium salts, sulfoxides and sulfones. However, such products are rare. They react electrophiles on carbon and hence resemble enamines, enols, ethers, and enol thioethers.

## Reactions on carbon

Five membered heteroaromatic ring systems having one heteroatom forms α-substituted product rather than β-substituted product due to effective **delocalization** of charge in the intermediate leading to α-substitution rather than intermediate forming β-substituted product.

In case of substituents present, those with -I, +M effects like halogens show α-directing effect. If the substituents is at 2-position, then C-5 substitution occurs. Substituents with -I, -M effects like $NO_2$, CN and COR; if substituent is at position-2 then a mixture of 4- and 5- substituted products shall be formed. C-4 is least deactivated by the substituent, but C-5 is most activated by the ring heteroatom.

    i.       Protonation

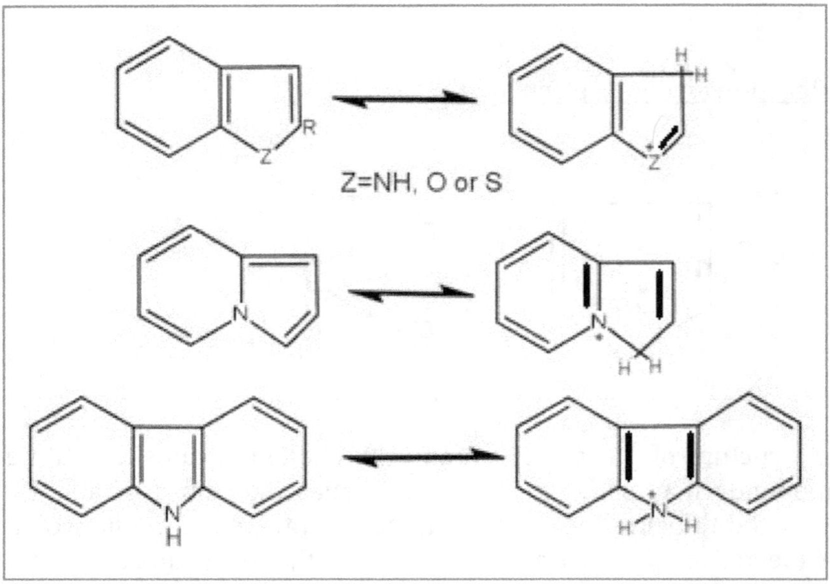

Stable α-protonated thiophenium ion results from the reaction of thiophene with hydrogen chloride and aluminium trichloride in inert solvent.

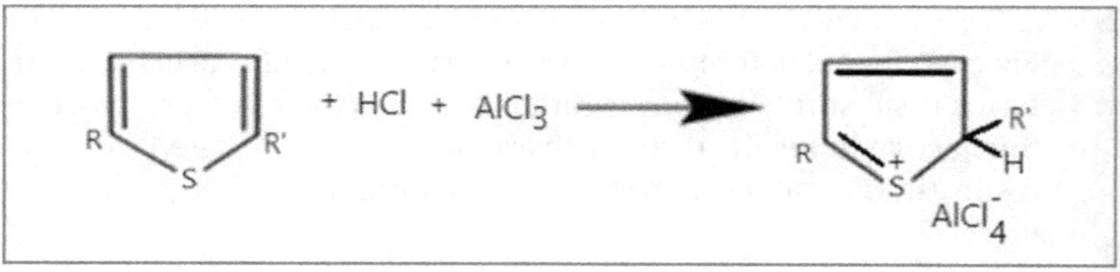

**Protonation** of pyrrole, furan and thiophene derivative generates reactive electrophilic intermediates which participate in polymerization, rearrangement and ring opening reactions.

ii.      Nitration

2-Nitro derivative along with some 3-nitro compound is formed when pyrrole is subjected to acetyl nitrate (obtained from fuming nitric acid and acetic anhydride) as nitrating agent. If α-position is substituted by electron withdrawing group, **nitration** of N- substituted pyrrole yields β- substituted products. Furan gives 2-nitrofuran under such condition.

Indole can be nitrated to 3-nitroindole at low temperature. 2-Methyl indole gives 3-nitro derivatives with nitric acid while 2-methyl-5-nitroindole with a mixture of concentrated sulphuric and nitric acid. Indolizines are also readily nitrated.

2-methyl indolizine                                    2-methyl 1-nitro indolizine

iii.    Halogenation

Five membered heterocyclic compounds with nitrogen, sulphur and oxygen as heteroatom are easily halogenated. **Bromination** of pyrrole with bromine in acetic acid gives 2,3,4,5-tetrabromopyrrole and iodination with iodine in potassium iodide(aqueous) results in tetra-iodo product.

Polyhalogenated products are formed at room temperature by reaction of furan with halogens. However, at low temperature, substitution occurs at position-2. **Halogenation** of thiophene also gives preferential results at $\alpha$-position.

iv.    Acylation

2-Acetyl and 2,5 – diacetyl pyrrole is obtained on heating pyrrole with acetic anhydride at 150-200°C. Acylation of either pyrrole or thiophene with chloro sulphonyl isocyanate gives 2- substituted amide which fragments on heating.

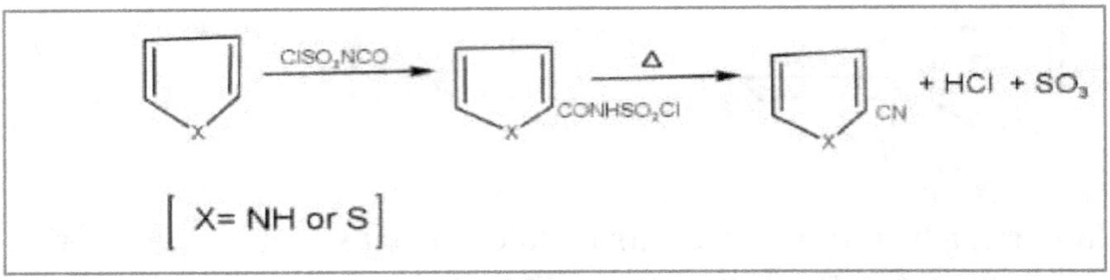

v.      Alkylation

Reaction of indole with excess of methyl iodide at 110°C gives tetra-methyl derivative via intermediate 2,3-dimethyl indole.

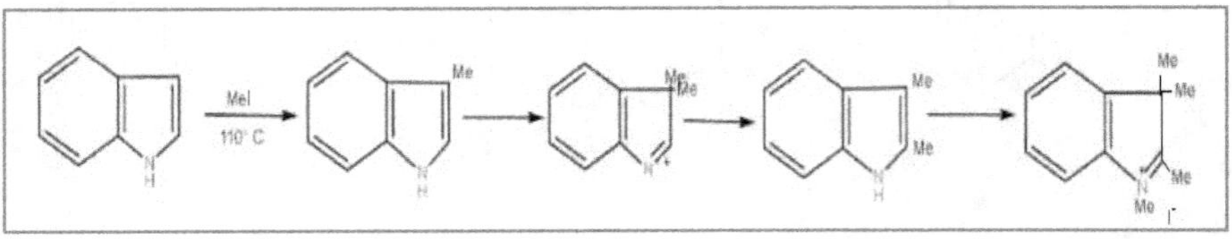

vi.      Diazo coupling

Pyrrole and indole give various products of this category.

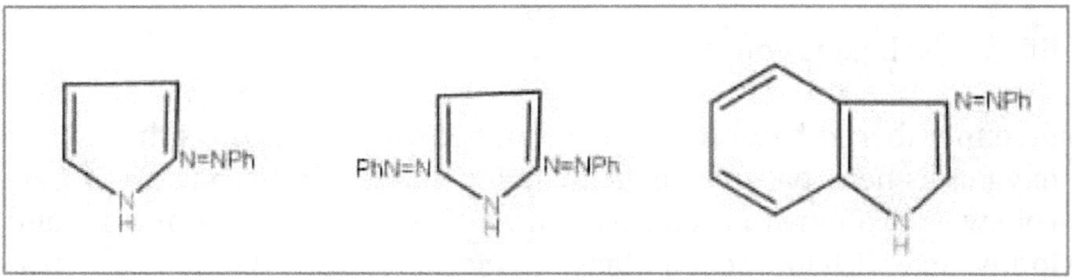

vii.      Nitrosation

Nitrosation of pyrroles or alkyl pyrroles results in either ring opening or oxidation of ring as is shown by the formation of meleimide from 2,3,4- trimethylpyrrole.

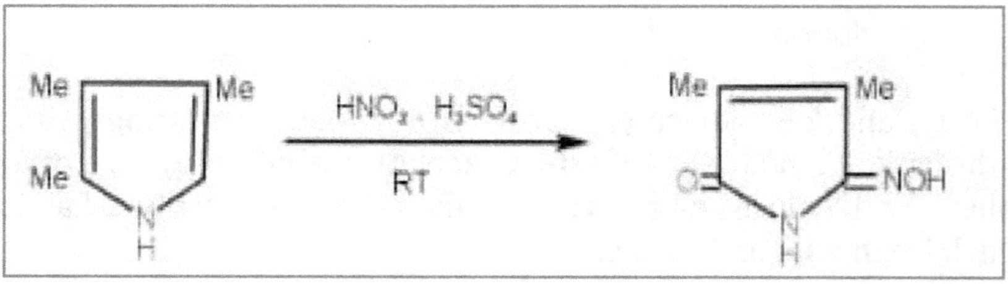

## Reaction on ring hetero atom

Alkylating agents form thiophenium salts conveniently with trimethyloxonium tetra fluoroborate ($M_3O^+BF_4^-$).

### i. Reaction with oxidizing agents

Pyrroles are easily reacted with oxidizing agents and give fragmented products.

### ii. Reaction with Nucleophiles

Five membered heterocyclic compounds containing nitrogen, sulphur and oxygen do not react with nucleophilic reagents by substitution or addition but only by proton transfer.

### iii. Reaction with reducing agents

Selective reduction in benzene ring of indoles can be achieved by treatment with lithium in liquid ammonia and selective reduction of

92

pyrrole ring can be got with sodium cyano-borohyride and acetic acid or triethylamine borane and hydrochloric acid.

## 6.4 PYRROLE & ITS BENZO DERIVATIVE

Bond length and angles determined from the analyses of microwave spectra are given in figure.

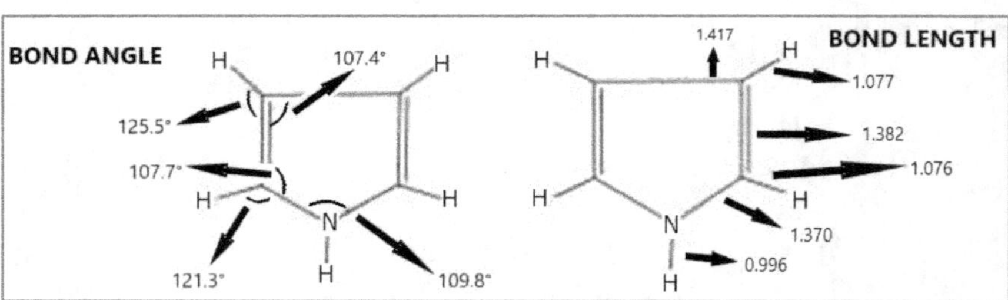


The N-C2 & C3-C4 bonds are shorter than normal single bonds whereas the C2-C3 bonds are longer than normal double bonds. The geometry reflects the **cyclic delocalization** implied by the canonical structure.

## 6.4.1 SYNTHESIS

The **reductive cyclization**, which happens to be general indole synthesis, proceeds by nitro group attached to aromatic ring as the most favoured precursor.

Starting from o – nitro benzaldehyde and nitromethane and proceeding via **partial reduction** also gives the required product.

The **Knorr–pyrrole synthesis** (cyclative condensation) proceeds by reaction between two components; an α-amino ketone furnishes a nucleophilic nitrogen and an electrophilic carbonyl thereby forming an enamine initially which cyclizes further.

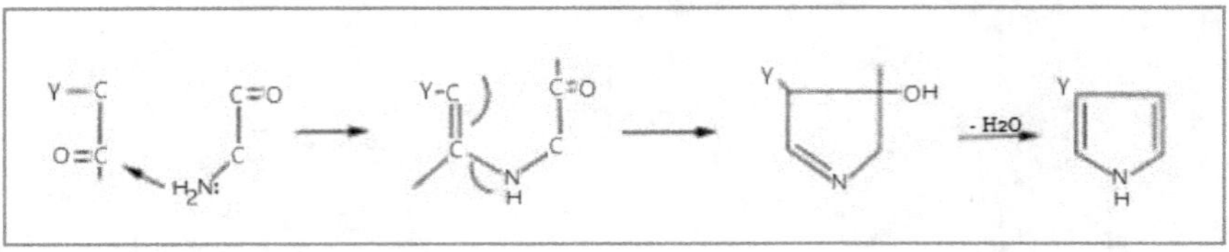

However, two alternative routes follow for cyclization when α–amino ketone and acyl/alkoxy carbonyl substituent are used as combination.

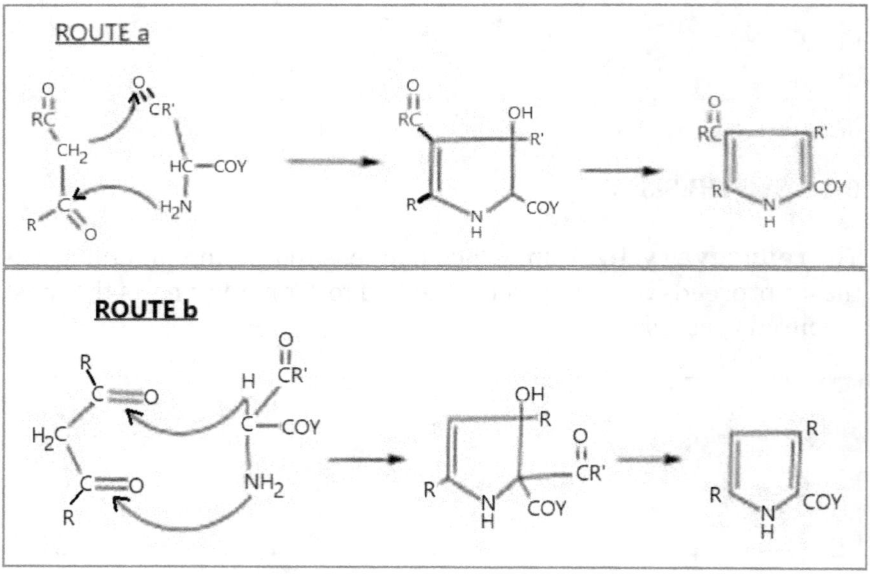

The **Fischer indole synthesis** involves the conversion of an N – arylhydrazone to an indole with the elimination of ammonia.

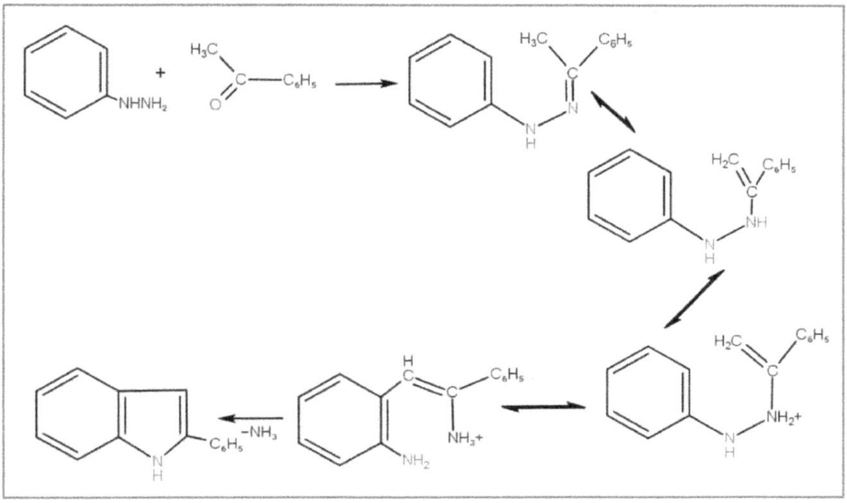

The method of synthesis in which nitrogen is not attached to benzene ring (**Nenitzescu indole synthesis**) leads to preparation of hydroxy indole from quinone and β – alkoxy carbonyl enamine.

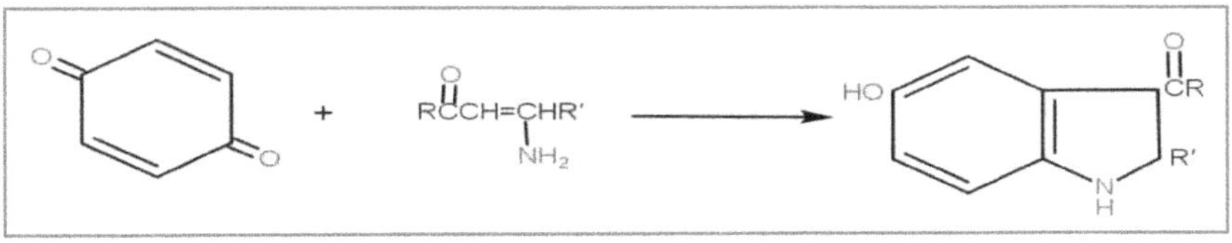

With unsymmetrically substituted quinones, the following table must be referred:

| | | |
|---|---|---|
| (1) 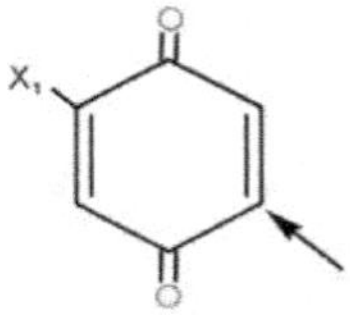 | $X_1$ electron donating group eg. OMe, OH | $X_1$ become C-6 substituent in the indole ring. C-3 of quinone deactivate for nucleophilic attack |
| (2) 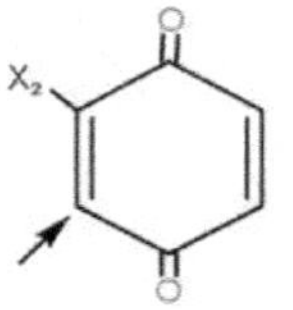 | $X_2$-electron atracting group eg. $CF_3$, $CO_2Me$ | $X_2$ become C-2 subsituent in indoid ring. C-3 of quinone activated to the nucleophilic attack |
| (3) 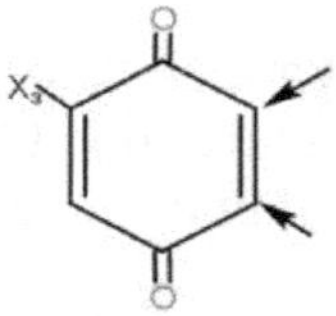 | $X_3$- Me, helogen | $X_3$ become either C-6/C-5 substituent |

**Photolytic ring contraction** of 3 – diazo – 3, 4–dihydro–4–quinolones also give indoles. Wolff Rearrangement as shown also results in indole formation.

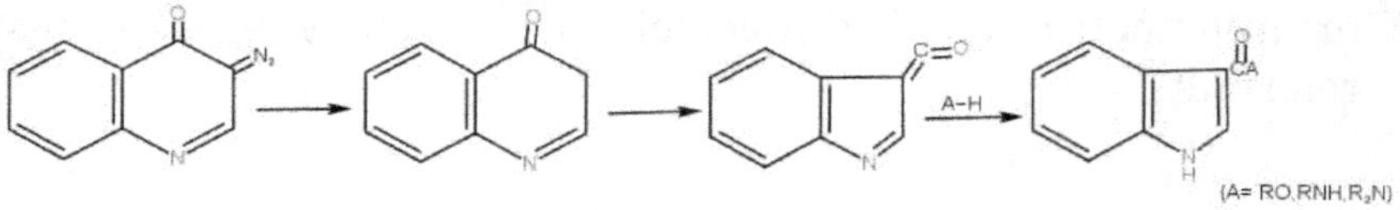

There are rare reactions of rearrangements of 2–acyl aziridines to yield pyrrole.

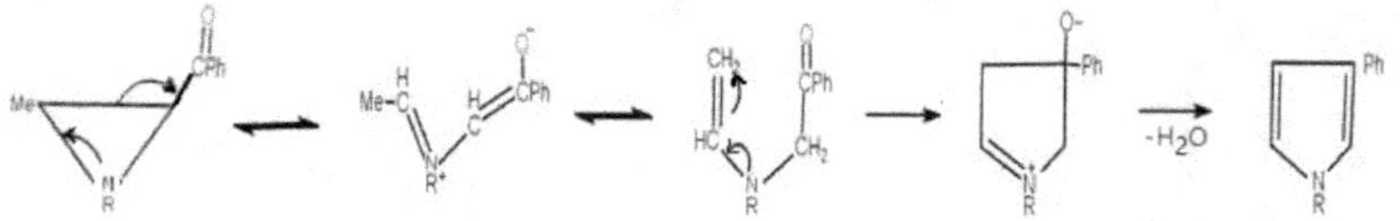

As an alternative route from pyrrole to indole; [4 + 2] cycloaddition results in dihydro or tetrahydro indole.

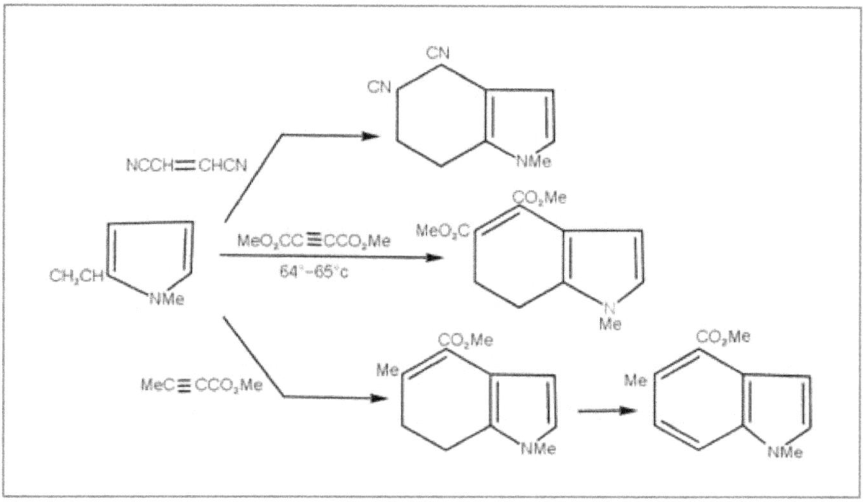

Indigo dye is prepared as indoxyls are extremely sensitive to air and on oxidation yield it as product.

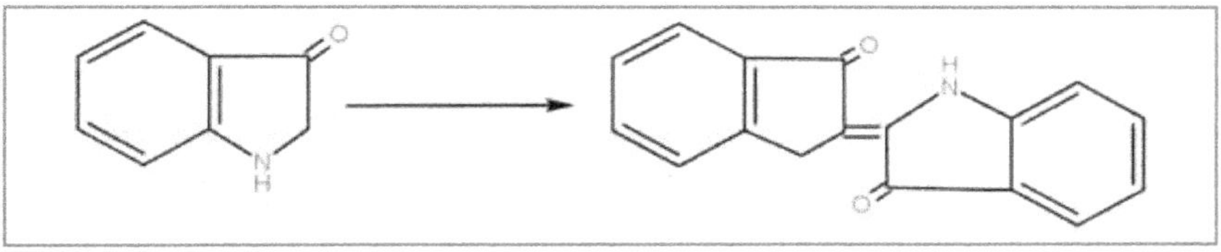

## 6.4.2 REACTIONS

The $\pi$- electron excessive character of pyrrole and indole renders both systems extremely susceptible to electrophilic attack.

### A) **Protonation**

3 – Protonated indoles have been shown to predominate at equilibrium in dilute acidic media, but protonation at the 1- and 2- positions also occurs in strongly acidic media. The substituents that are **electron – withdrawing** usually reduce the basic tendency of the heterocyclic rings. There is ample evidence (spectroscopic) showing that acylpyrroles are protonated at carbonyl oxygen while $\alpha$ – protonation is minimum in the ring.

Protonation of pyrroles and indoles enhances **nucleophilic addition** to the ring. Hence, they both give results with sodium hydrogen sulphite thereby forming adducts. Protonated pyrrole ring when brought in contact with hydroxyl amine or phenyl hydrazine results in cleavage.

## B) <u>Nitration</u>

The resultant products of reaction with nitrating agents depend on reaction conditions. When the agent is nitrous acid, a complex mixture of products is obtained for indole as 3-oximino $3H$ – indole along with dimeric products.

Pyrrole may be nitrated by nitric acid at low temperatures in acetic anhydride to yield 2 – nitro pyrrole (major product) whereas indole, reacts with ethyl nitrate in the presence of sodium ethoxide to give 3- nitro indole. If there is an **electron withdrawing substituent** at 3- position of indole ring, then a variety of products are obtained under different reaction conditions.

## C) <u>Halogenation</u>

Chlorination and bromination have been reported using varied reagents although there are fewer examples of fluorination reactions. Chlorination of indole with sodium hypochlorite results in the formation of 3 – chloro indole which further yields 3, 3 – dichloro indole.



When indole reacts with sulfuryl chloride, 3- chloro indole is obtained, which in excess of sulfuryl chloride, gives 2, 3- dichloro indole. The presence of an **election – withdrawing substituent** on the five ring of the indole deactivates that ring to electrophilic attack by both elemental bromine and N – bromo succinimide.

100

### D) <u>Acylation</u>

The **Friedel Craft acylation** of pyrroles and indoles give good results when the reactants are substituted with electron – withdrawing groups.  Unsubstituted pyrroles and indoles react with acetic anhydride in the absence of a Lewis acid to give 1-acetyl and 1,2 – diacetyl pyrrole together with 1, 3 – diacetyl indole as a minor product and 1-acetyl and 1,3-diacetyl indole. At high temperatures, 1,2 – diacetyl pyrrole rearranges to give 2, 5 – diacetyl pyrrole.

 Friedel crafts acylation of 3 – substituted indoles give 2 – acetyl derivatives in very vigorous conditions than those required for acylation at the 3- position and probably proceeds by rearrangement of 3 – acetyl -3- substituted 3 – H – indolium cation.  Formyl pyrroles have been obtained from **acid catalysed reaction** of the pyrrole with ethyl ortho-formate. The same reaction of indoles leads to oligomerization. The reaction of indolyl anion with ethyl formate yields 1-formylindole at low temperature and 3-formyl derivative at high temperature.

### E) <u>Alkylation and Arylation</u>

The heteroaryl anion is attacked in direct alkylation of pyrroles and benzo pyrroles with haloalkanes. The Friedel Craft alkylation gives low yields at low temperatures. The pyrroles which are substituent by **election – withdrawing groups** show complications by the co – ordination of the substituent with Friedel – Craft catalyst. Direct alkylation of pyrrole nucleus using allylic and benzylic halides alkylates pyrrole at elevated temperatures. Indoles also react directly with allylic halides, while benzyl halides, unless they are substituted at the o-or p-position by election donating groups, are insufficiently reactive to benzylated indole. Methyl iodide reacts with indole only at temperature above 100°C. Under these circumstances, poly-methylation of both the compounds occurs.

excess MeI at 100°C

Alkylation of substituted indoles (2,3-disubstituted) leads to 2,3,3-trisubstituted indole but if the six – membered ring is activated to electrophilic attack, **Friedel – Craft alkylation** takes place at 5- or 6-positions. **Michael adducts** are got from the reaction of pyrroles with ethyl propiolate and But-1-yn -3- one. Indoles react with propiolic acid to form 1,1-bis (3-indolyl) ethane via the 1:1 Michael adduct.

## F) <u>Reaction with aldehydes and ketones</u>

The reaction of pyrroles and indoles with aliphatic aldehydes are similar to those involving aromatic aldehydes yielding the unstable aza fulvenium salts along with the pyrroles – methane or indolyl – methane. 2,5 - and 2,3 – Disubstituted pyrroles from 2:2 adducts (A, B, C) and the macrocyclic compound (D) is obtained from the reaction of formaldehyde with ethyl 2 – methyl pyrrole – 3 – carboxylate.

In the same way, indole reacts with metaldehyde and forms adducts.

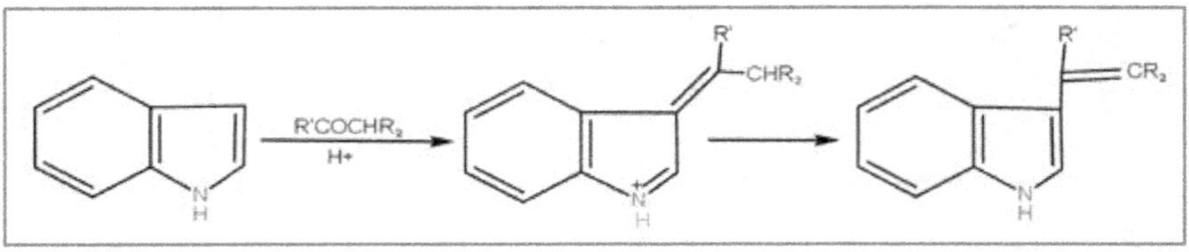

The **reductive alkylation** of the pyrrole ring using a range of aliphatic and aromatic aldehydes and ketones, may also be accomplished with phosphonium iodide, with hydrochloric acid and zinc amalgam, or with tin (II) bromide in hydrobromic acid.

## G) <u>Reaction with Nucleophiles</u>

The $\pi$ – election excessive property of pyrrole and benzo pyrrole makes them unreactive for **nucleophilic reactions**. It is for this reason that such reactions are rare and only take place when electron – withdrawing substituents are present.

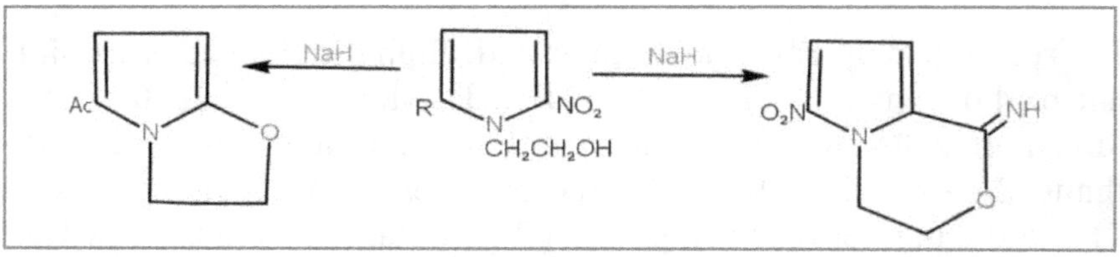

1, 4 – Nucleophilic addition reaction occurs when 3- Benzoyl – 1,2 – dimethyl indole reacts with phenyl magnesium bromide and indoline is obtained as product.

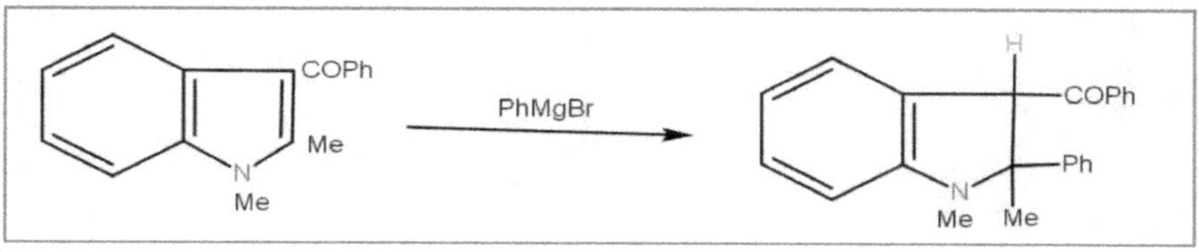

## H) <u>Oxidation</u>

Coloured polymeric products are obtained when pyrroles and iso indoles experience ariel oxidation however indoles are less readily oxidized by air & produce monomers or dimers and trimers.

## 6.4.3 APPLICATIONS

The practical applications of pyrroles and indoles is focussed on pharmaceutical field. The heterocyclic compounds in this category provide medicinal uses of two types:- anti – inflammation drugs and antihypertensive effects.

### 1. INDOMETHACIN

(i) anti – inflammatory agent in the treatment of osteoarthritis and gout.

(ii) used in rheumatoid arthritis.

(iii) as a prostaglandin synthetase inhibitor.

104

Two pyrrole derivatives, **Tolmetin** and **Clopirac** which have structure resemblance to indomethacin also find use in medicine industry.

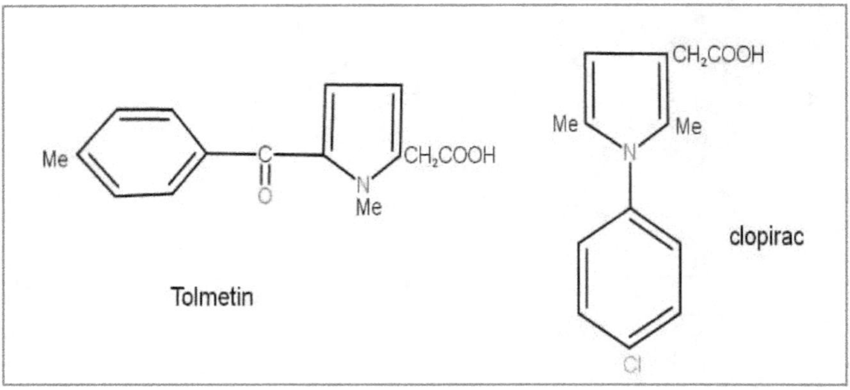

2.   **SEROTONIN** is of use in the central nervous system and a vasoconstriction.

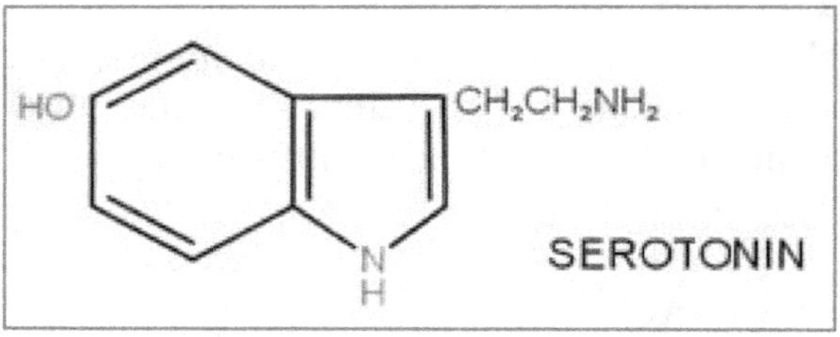

3.   **OXYPERTINE** finds use as tranquilizer, **INDORAMINE** as a hypertensive drug and **IPRINDOLE, TANDAMINE** and imipramine as antidepressants.

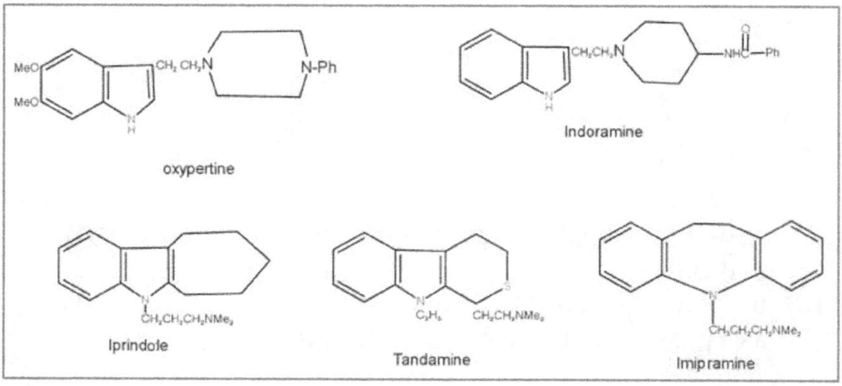

4. These are various pyrroles and indoles natural products.

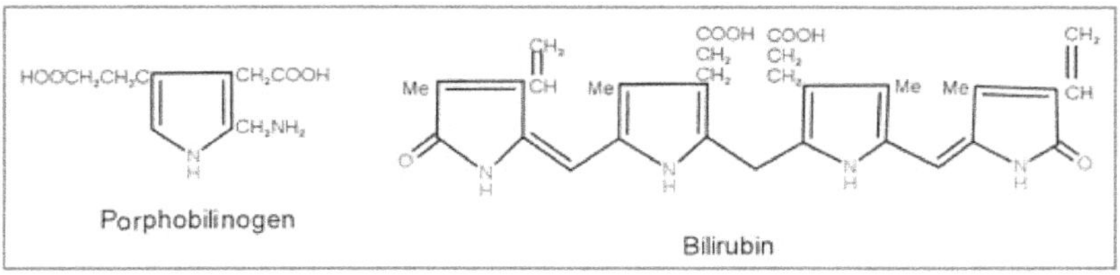

Porphobilinogen

Bilirubin

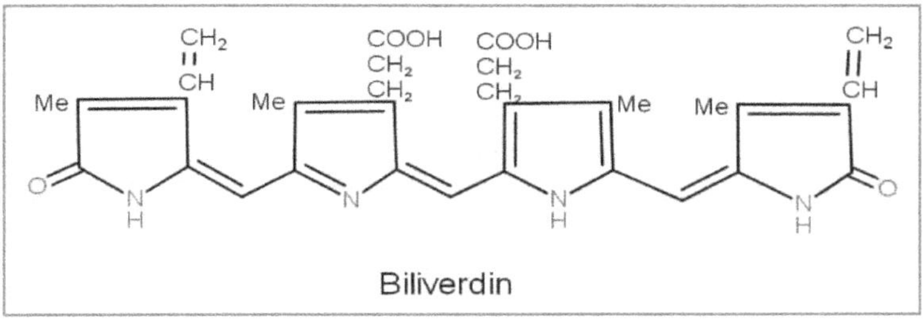

Biliverdin

5. **PYRROLNITRIN** is a good antibiotic against various microorganisms. **MITOMYCIN C** gives good results in treatment of cancers.

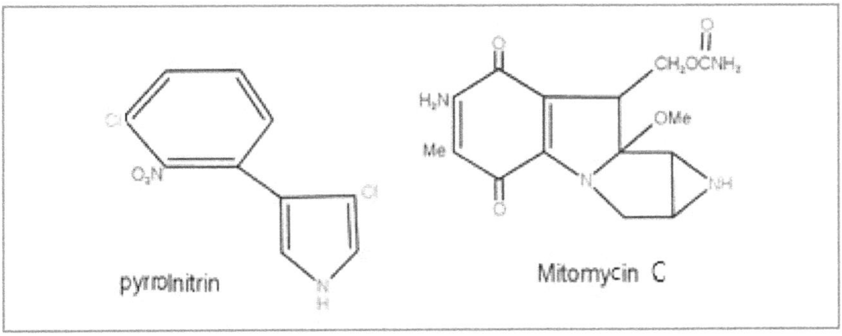

pyrrolnitrin

Mitomycin C

## 6.4.4 MACROMOLECULES CONTAINING NITROGEN

Large number of extremely important natural products have porphyrins and corrins as a part of prosthetic groups; however, phthalocyanines have found usage as dyestuff. Porphin is the parent macromolecule for porphyrin containing four pyrroles linked by four methine bridges.

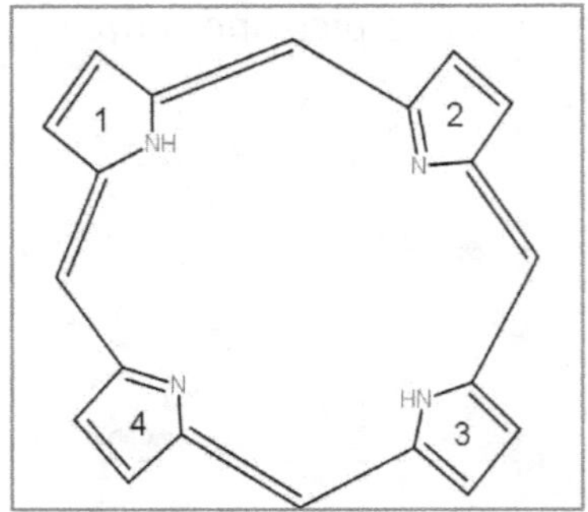

**PORPHIN**

Chlorin, which is the major constituent of chlorophyll is structurally like porphin with two hydrogens added to pyrrole ring, (4).

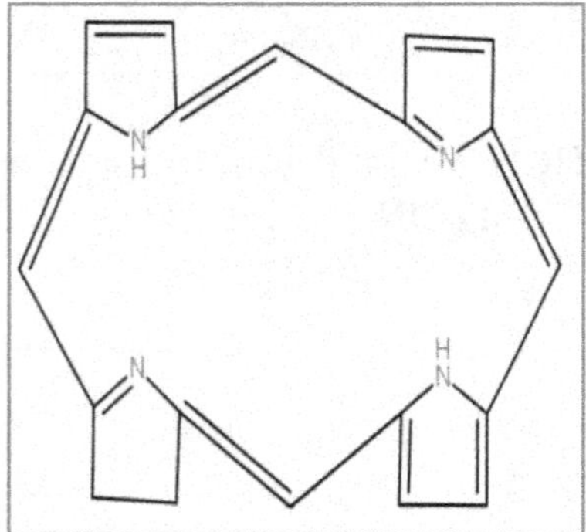

**CHLORIN**

Hence, **A**, iron (II) complex of protoporphyrin-IX) is the prosthetic group of hemoglobin, myoglobin, catalases, peroxidases and cytochromes. Clorophyll C, constituent of marine algae etc, is a mixture of two porphyrin acrylate **B** and **C**.

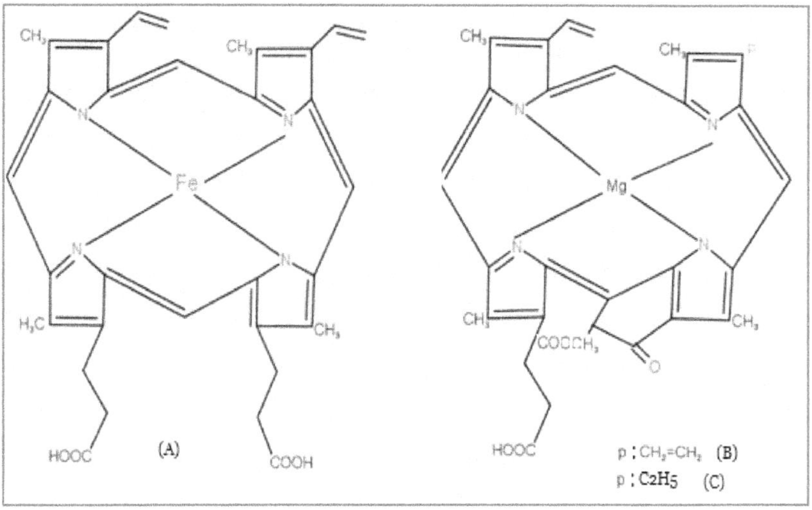

## Synthesis

The best synthesis is obtained by a series of steps. Step one is the synthesis of dipyrryl methene for which there are two approaches. **First** is to synthesize via 2,3-dimethyl pyrrole and 3,5-dimethyle pyrrole-2-aldehyde and **Second** is the formation of bromo-substituted dipyrryl methene.

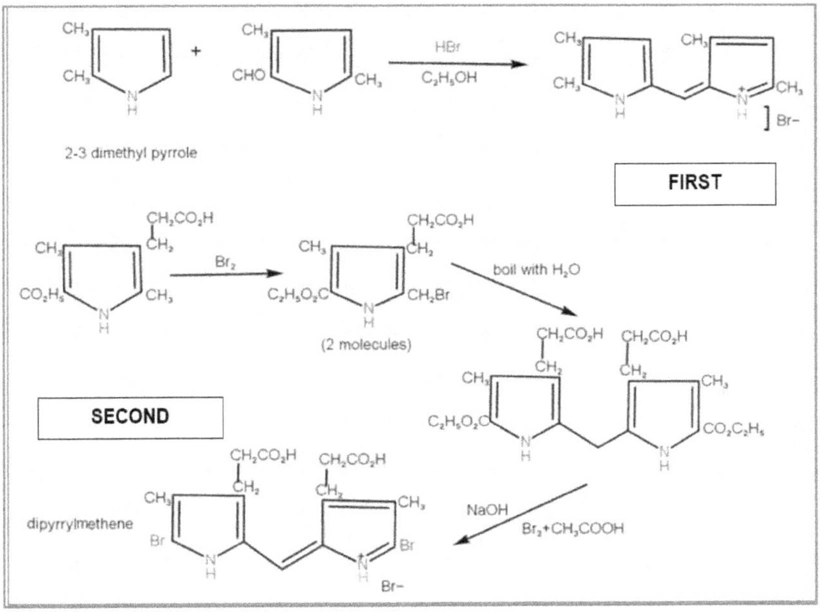

108

Then these two dipyrryl methene, synthesized in first and second approaches, condense in presence of succinic acid to yield deutero porphyrin which gets converted to deutero haemin which finally results in protoporphyrin.

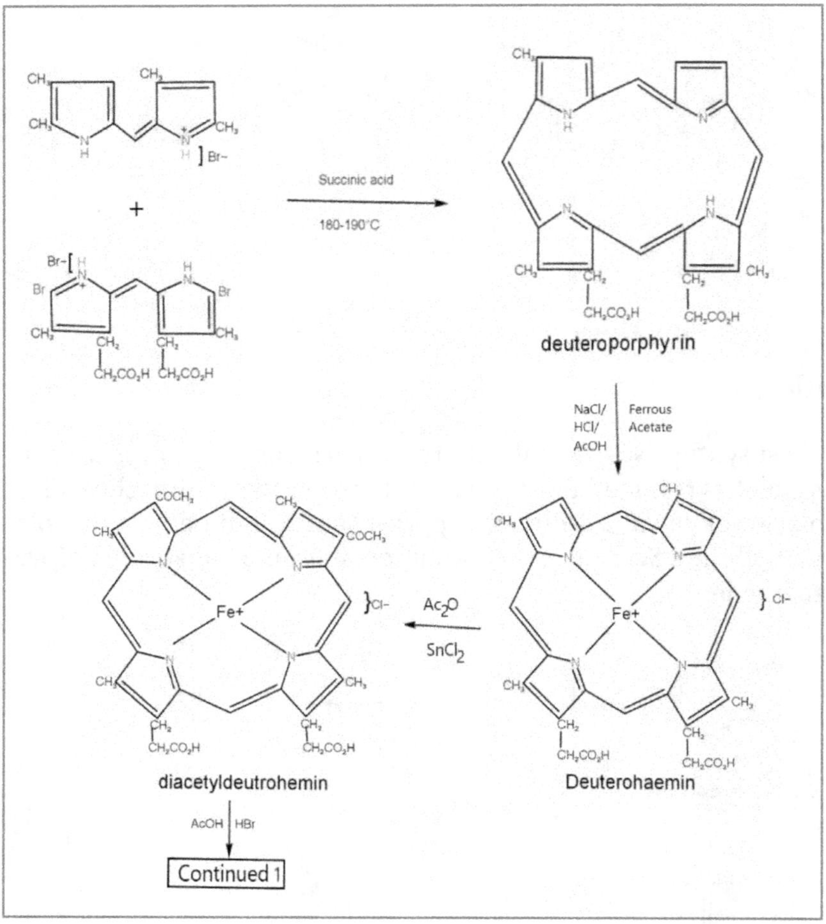

deuteroporphyrin

diacetyldeutrohemin

Deuterohaemin

Continued 1

Continued 1

Diacetyl deutroprophyrin

EtOH | KOH

Heematoprophyrin

distil at
105°C
in 25% HCl

Continued 2

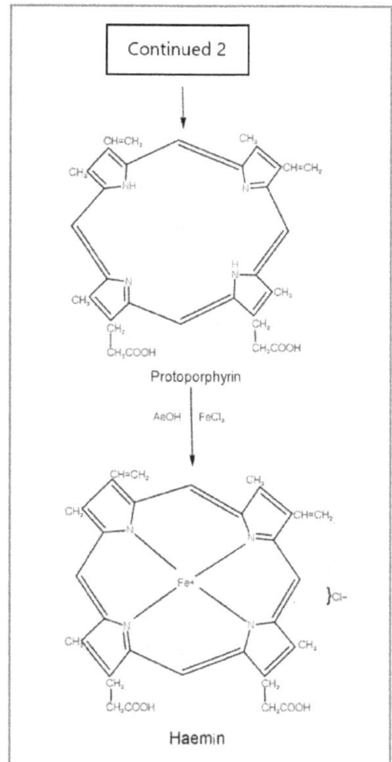


## Reactions

Oxidation of heme with hydrogen peroxide and ascorbic acid in pyridine results in oxidation of methine bridge. In acidic medium, iron oxophlorin exists in enol form which converts to keto form on treatment with base. On further exposure to oxygen, ring cleavage takes place and formation of biliverdin occurs.

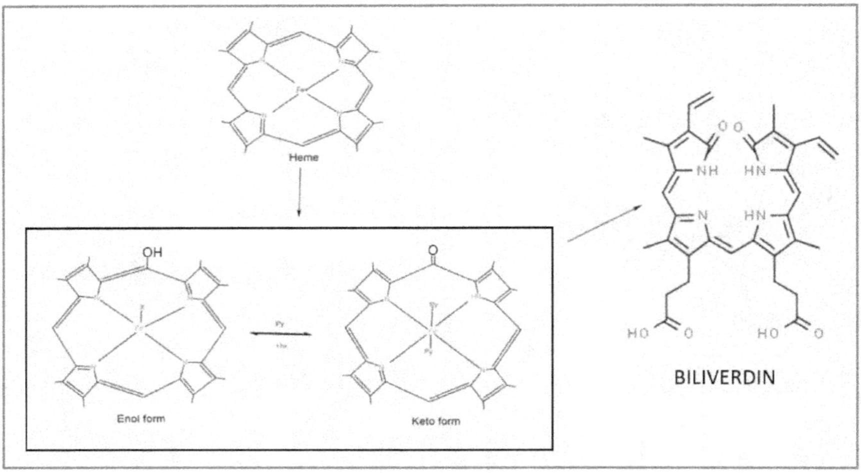

When porphyrins undergo oxidation with hydrogen peroxide and sulfuric acid, they form dihydroxy chloric which via **Pinacol Rearrangement** results in ketones.

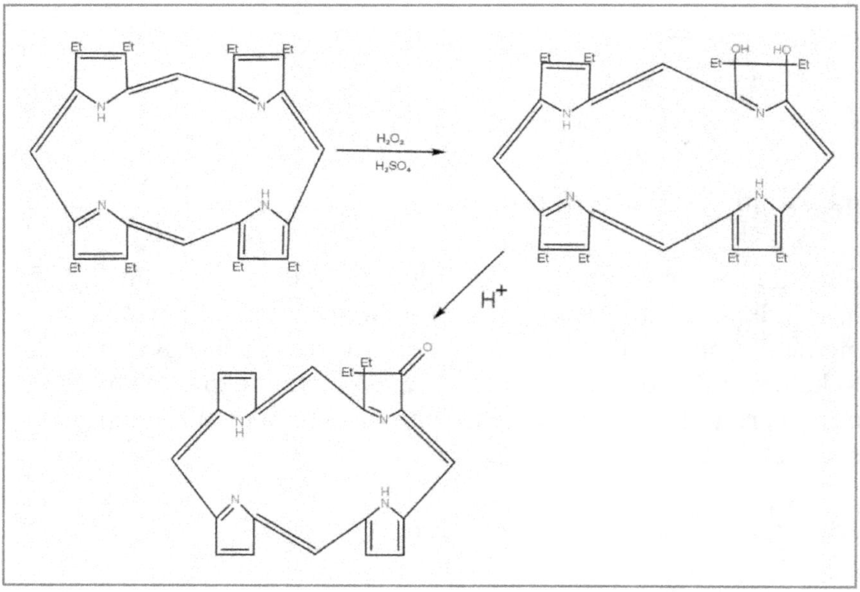

In the presence of chromium trioxide or potassium permanganate, degradation of porphyrin takes place. This reaction also serves the purpose of establishing the structure of the macromolecules.

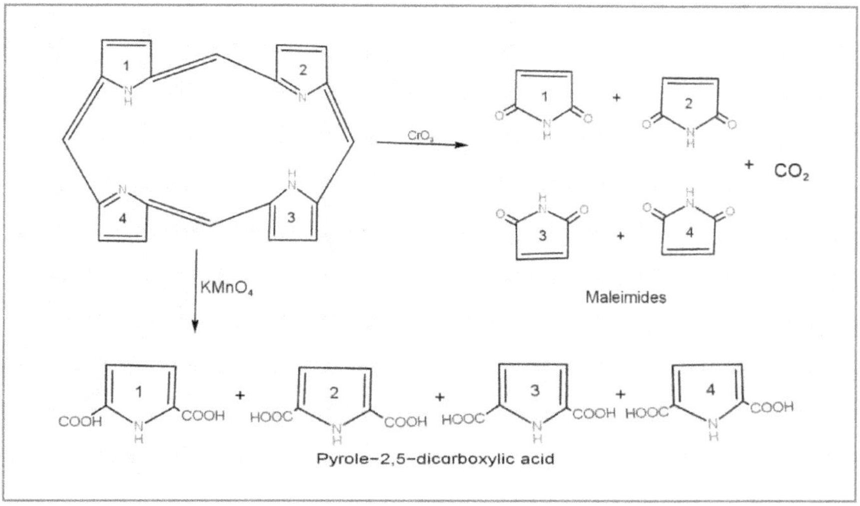

With the structural elucidation through $CrO_3$, there is a drawback that four bridges are lost as carbon monoxide; hence to overcome this protoporphyrin-IX (B) whose **oxidation** gave ethyl methyl maleimide (C) & hematinic acid (D).

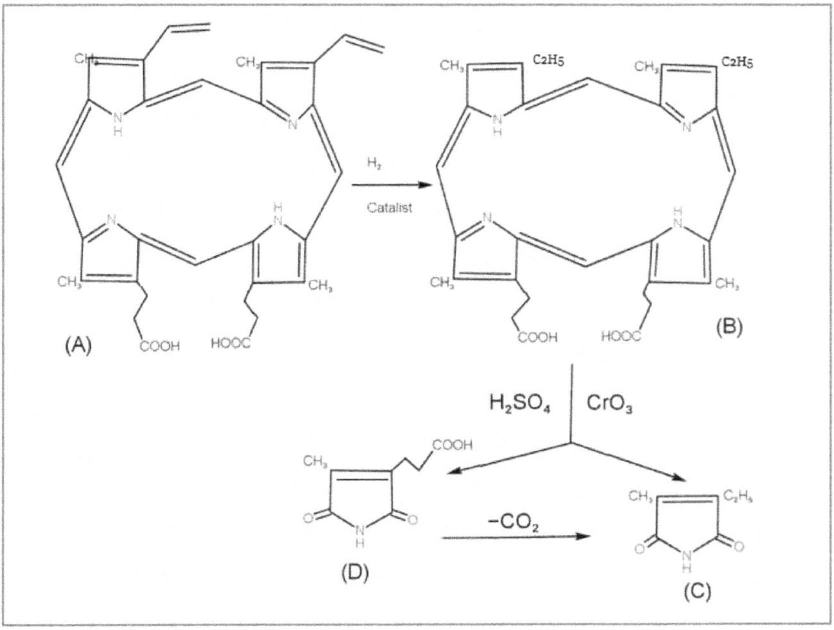

**Acetylation** reaction of porphyrin gives 3,8-diacetyl deuteron haemin-IX, when deutero haemin is treated with acetic anhydride in the presence of $SnCl_4$.

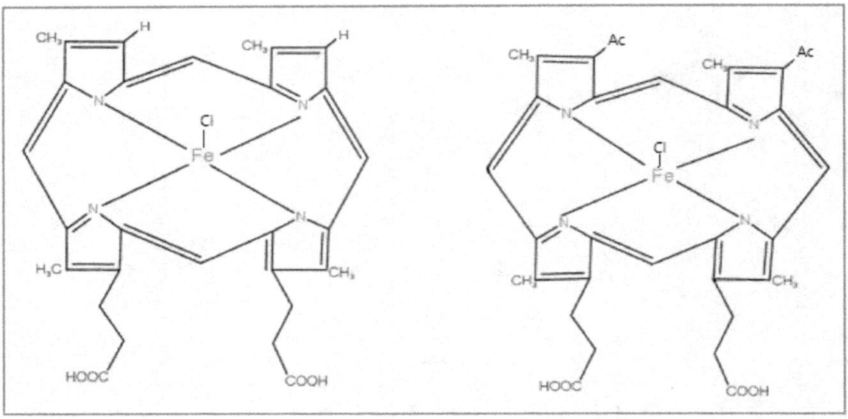

The propionic acid side chains of meso porphyrin-IX experience intramolecular acylation in the presence of sulfuric acid and results in isomeric forms.

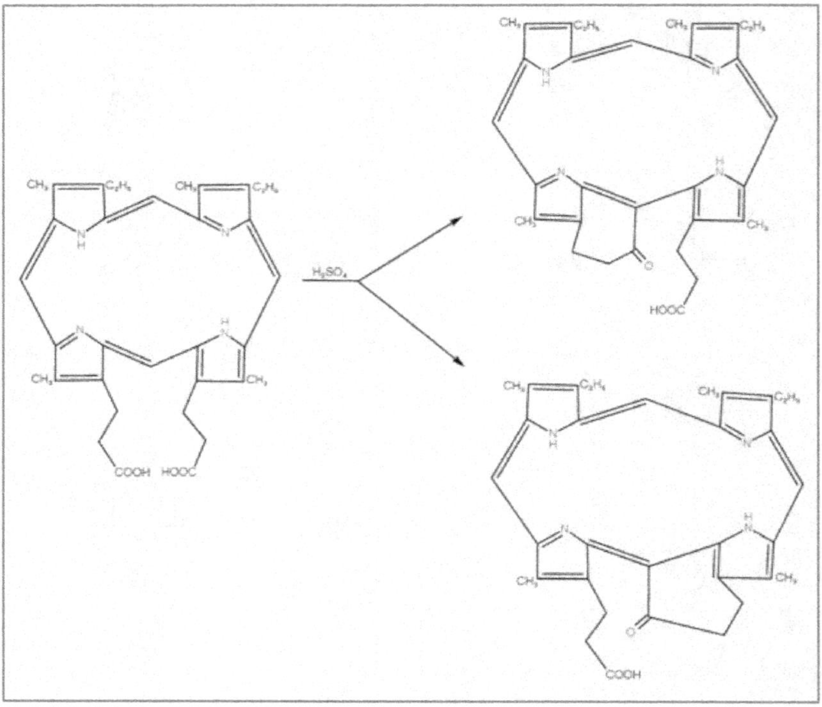

When chlorophyll-a is treated with alkali it results in the formation of pyro porphyrin-XV (A), phyllo porphyrin-XV (B) and rhodo-porphyrin-XV (C).

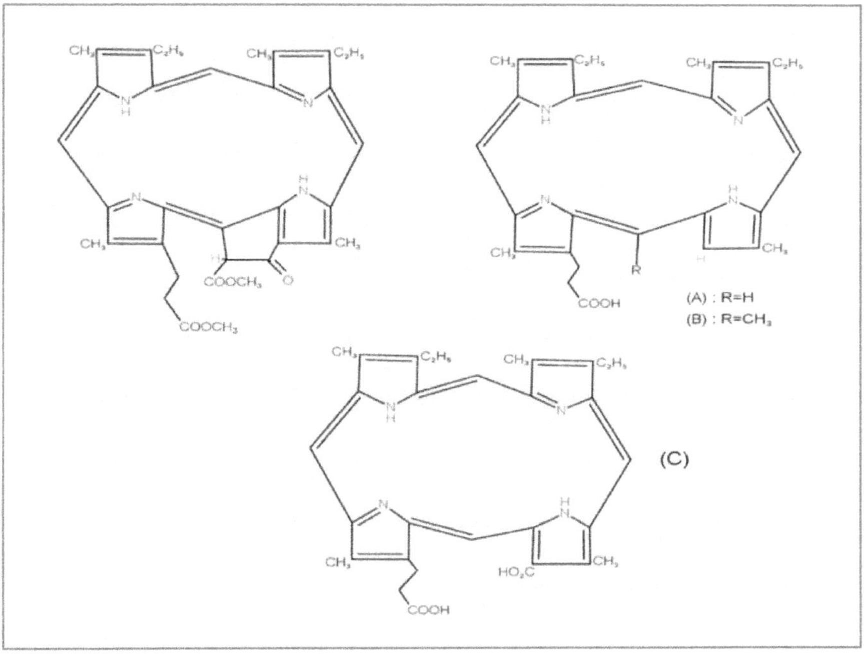

Phthalo-cyanins are tetra benzo-tetra aza-porphyrins. They are extremely stable to light and insoluble in most solvents hence they are extremely useful as dyes.

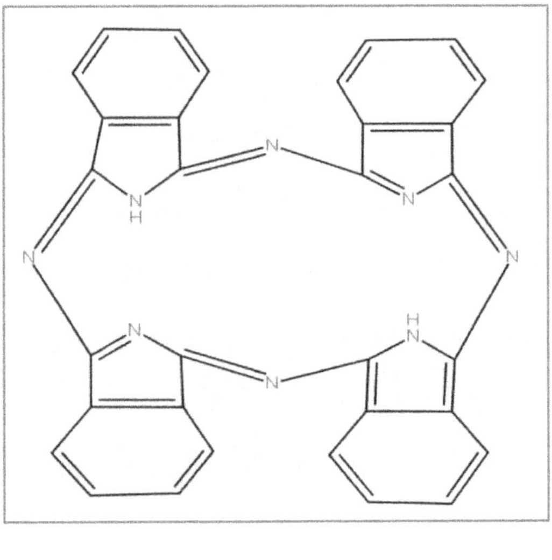

## 6.5 FURAN & ITS BENZO DERIVATIVES

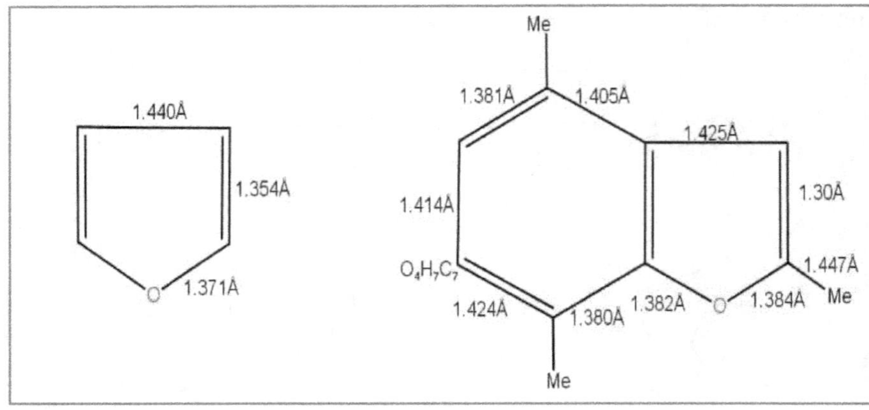

The bond lengths of Furan & its benzo derivatives are obtained by microwave techniques, X-ray or electron diffraction techniques.

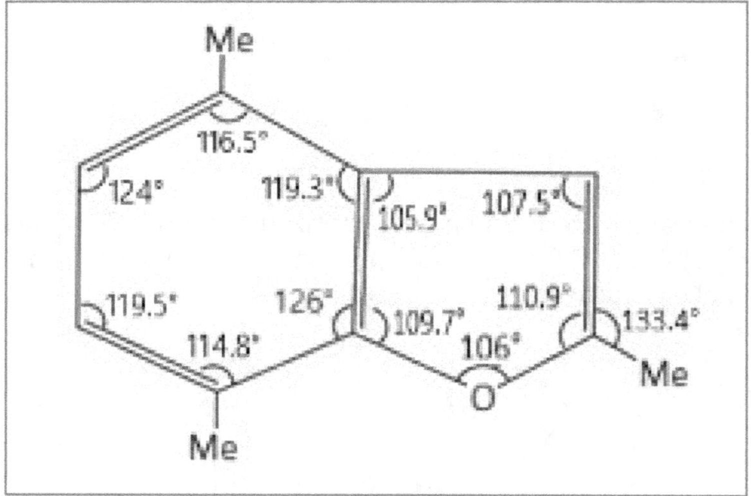

## 6.5.1 SYNTHESIS

The most studied preparation of Furan is by acid catalysed dehydration of 1,4-diketone followed by cyclization of the mono-enol. Furan is synthesised by cyclization of unsaturated 1,4-diketone in acetic anhydride and 3-acetoxy-2,4,5-triphenylfuran is obtained as product.

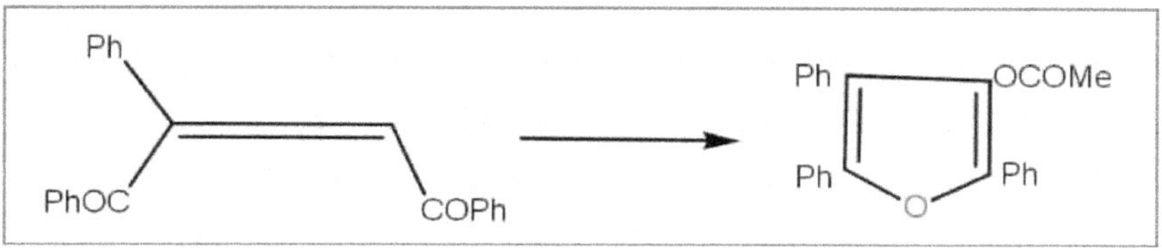

When β-aryl propionamide reacts with acetic anhydride and perchloric acid, a salt is obtained which on deprotonation with triethylamine gives 2-aryl-5-dialkyl amino furan.

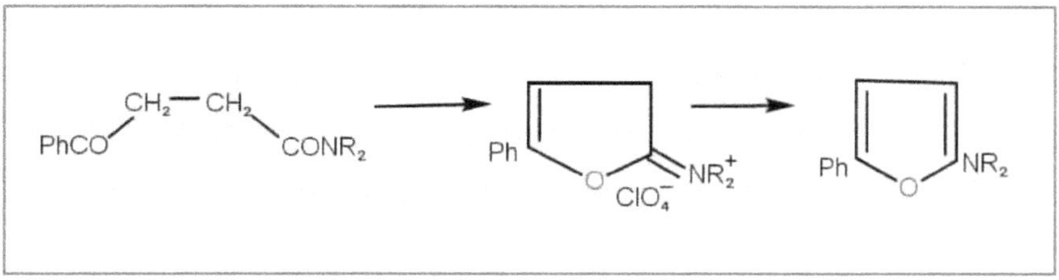

3-substituted furans are obtained when alkynic epoxides (product from chloro acetylenic carbinols) react with ethyl alcohol in sulfuric acid & mercuric sulphate.

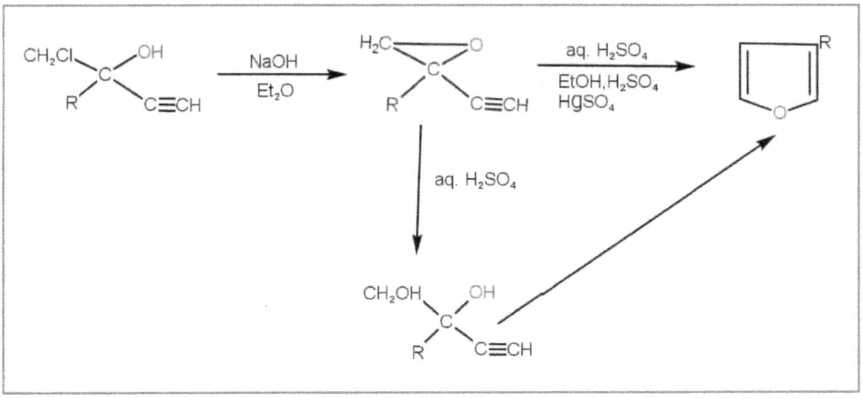

A novel method of preparation of acyl furan is shown by the formation of 3-acetyl derivative by heating bromoalkene.

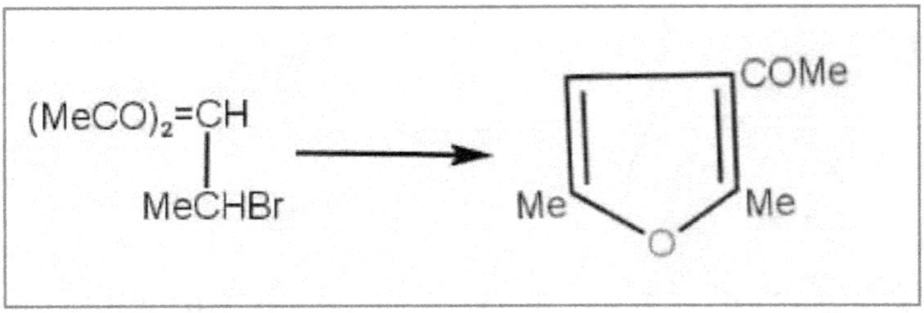

Dibenzofurans are obtained as minor products after cyclization (via **UV irradiation**) of diphenyl ethers (A) and (B) in the presence of Iodine and treatment of (A) with palladium (II) acetate in boiling acetic acid yield a mixture(C).

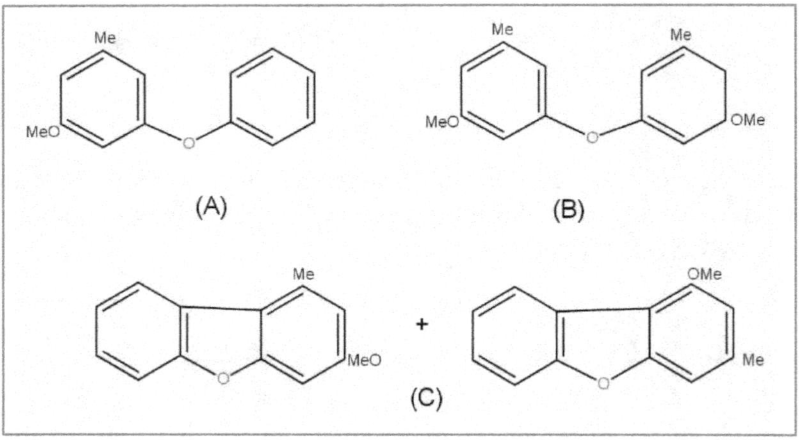

(A)

(B)

(C)

**Photochemical reduction** of o-dibenzoyl benzene is brought about in ethanol to produce 1,3-diphenyl benzo[c]furan. The same product is also obtained when o-dibenzoyl benzene is treated with zinc dust in acetic acid.

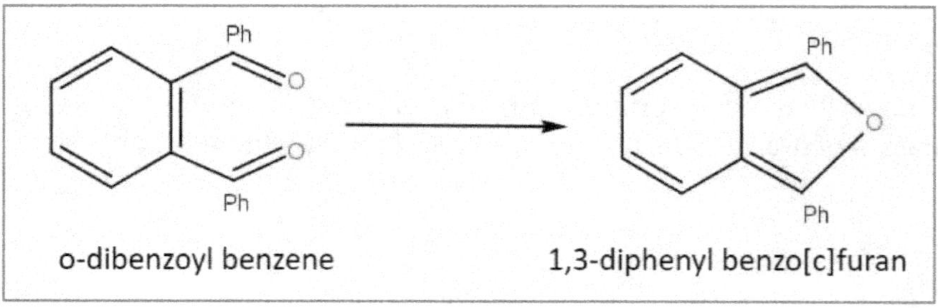

o-dibenzoyl benzene        1,3-diphenyl benzo[c]furan

When 1,4-diphenylbutadiene and dibenzoyl acetylene undergo **Diels-Alder reaction**, o-diaroyl benzene is obtained along with 1,3,4,7-tetraphenylbenzo[c] furan.

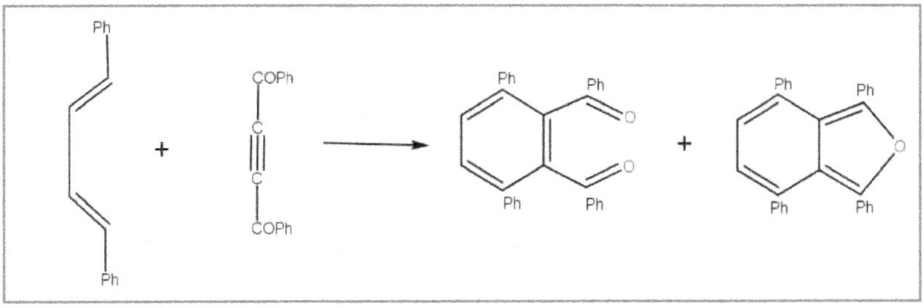

Substituted furans are produced when monosaccharides react with a variety of 1,3-dicarbonyl compounds in the presence of zinc chloride with ethanol or water as solvent.

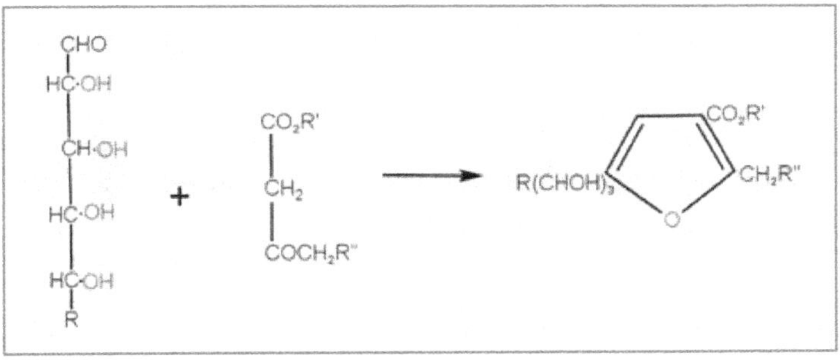

A reaction of 2-ethyl-3-methyl-4-pyrone in the presence of light and sulfuric acid gave 2-acetyl-5-ethylfuran. However, 2,6-dimethyl-4-pyrone also gave substituted furan along with substituted 2-pyrons.

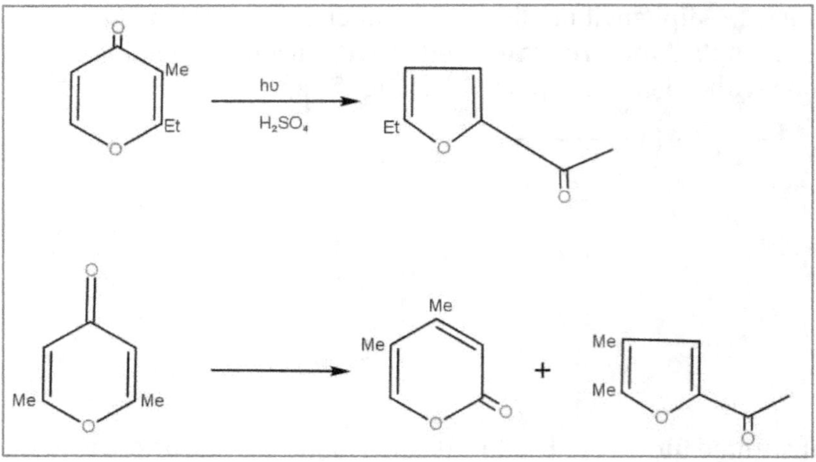

118

## 6.5.2 REACTIONS

## 1. Protonation

Protonation reaction can lead to ring opening; for example, reaction of 2,5- dimethyl furan with perchloric acid and aqueous DMSO (dimethyl sulphoxide) is one of this kind.

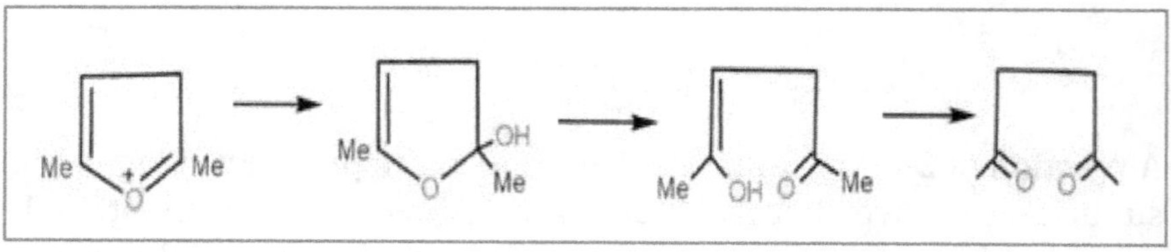

## 2. Halogenation

**Bromination** of furan in methanol at -10°C in the presence of a base gives 2,5- dimethoxy – 2,5- dihydrofuran. When bromination is brought about by election-withdrawing groups, 5- bromo -products are obtained.

## 3. Acylation

These set of reactions give but products with agents like acetic anhydride or boron trifluoride etherate or phosphoric acid. Tin (IV) chloride is also used as a good catalyst.

## 4. Reaction with diazonium salt

When substituted furan reacts with 2,4- dinitrobenzene diazonium sulphate (in aqueous acetic acid), azo compound is obtained (A); however, with 4-nitrobenzene diazonium chloride in aqueous ethanol & sodium acetate, an intermediate product (B) is obtained which converts into pyrazole (C) by hydrochloric acid.

## 5. Reaction with Nucleophiles

There are a variety of reactions in this category, some of which have been listed below.

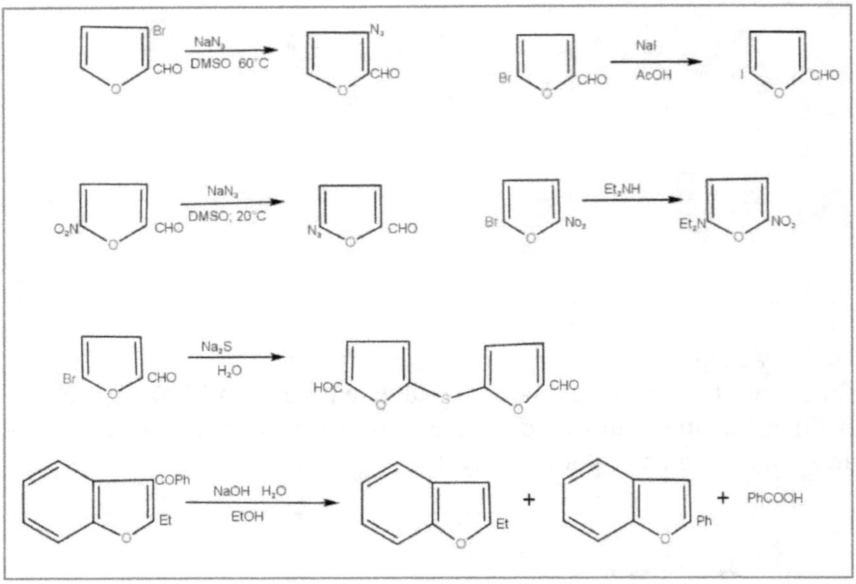

## 6. Reaction with reducing agents

Reduction of benzofuran takes place in the effect of various agents.

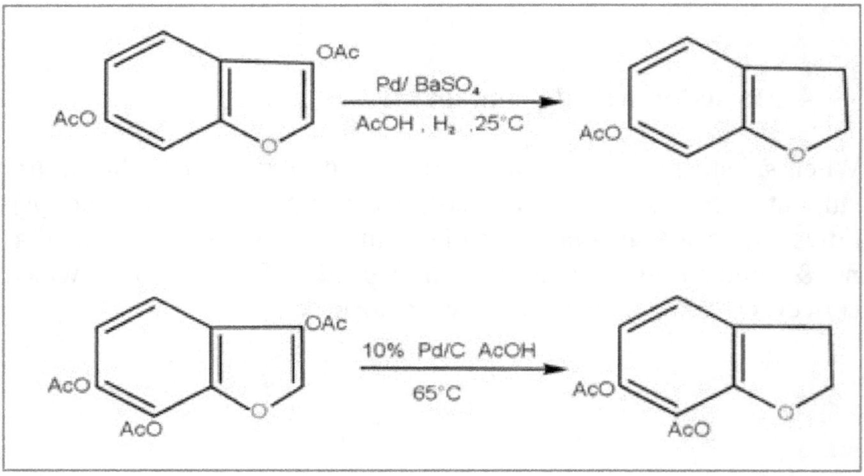

## 7. Reaction with free radicals

Substituted products are obtained if methyl free radical is generated from diacetyl peroxide.

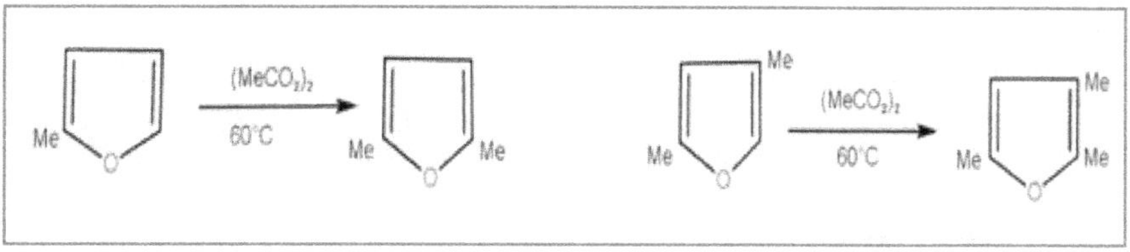

## 8. Cycloaddition reaction

Cycloaddition in furans gives products in good yield.

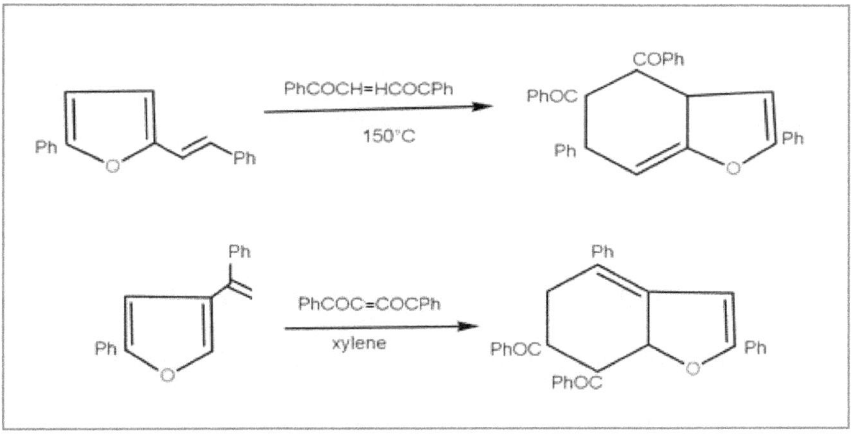

The enhanced aromaticity of benzofurans prevents it from taking part in Diels-Alder Reaction; however, 2-vinyl benzo furans give adducts.

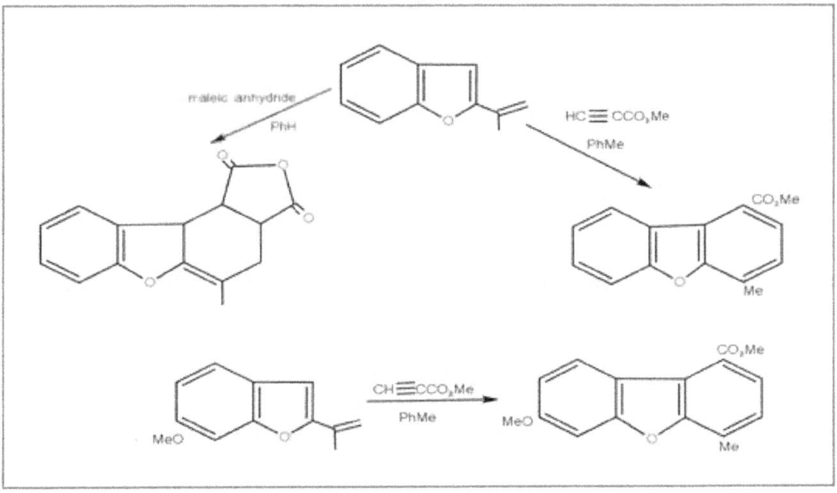

**Photodimerization** is not a property of furans; however, benzofurans undergo this reaction giving low yields; but oxidative photocyclization occurs in furans which are a part of hexatriene system.

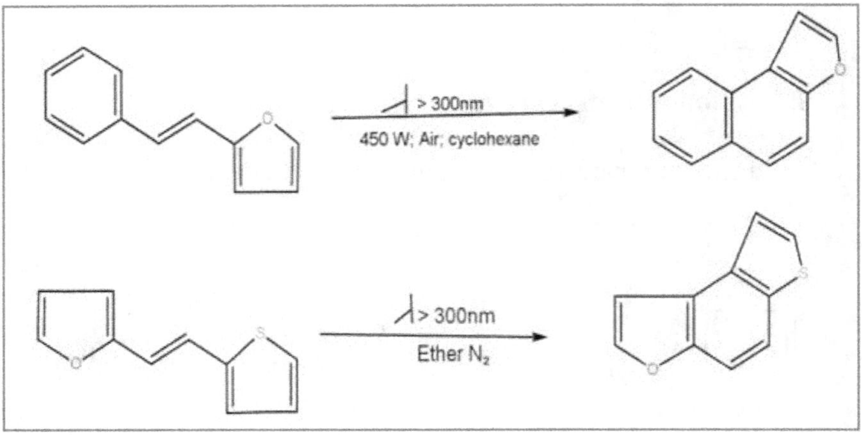

### 6.5.3 APPLICATIONS

Of all the furans, furfural has great industrial importance as it is widely used in plastics, rubber, petroleum, synthesis of furans and vegetable oils. **Furaneol** is present in pineapples and strawberries. The $\alpha$- methylene- $\gamma$- butyrolactone (A) group is contained in many terpenes.

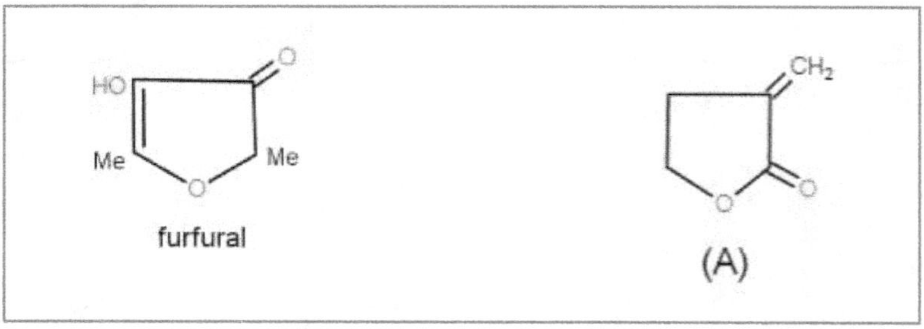

furfural

(A)

The **tetrahydrofuran** derivatives (B) and (C) are antidepressants.

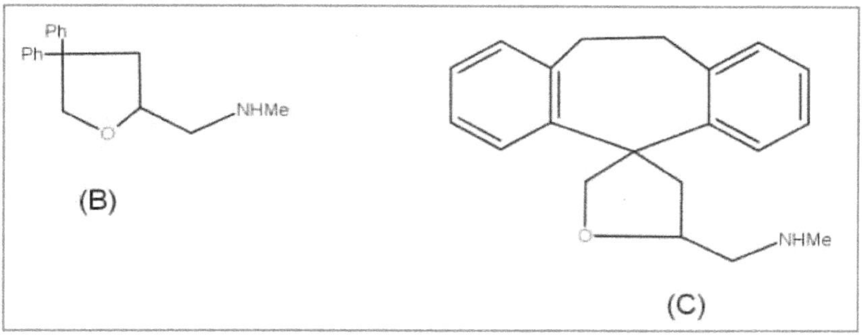

(B) (C)

The **trimethylpsoralen** (D) and its derivative (E) have potential to act as photoreactive crosslinking reagents for nucleic acids.

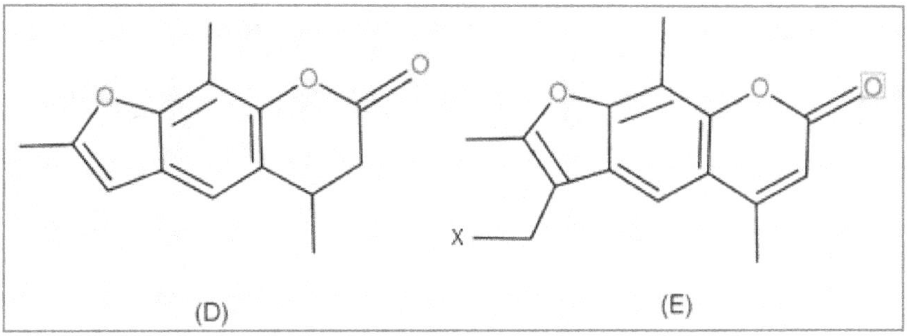

(D) (E)

Furocoumarins such as **psoralen** (F) and **methoxalen** (G) which are naturally occurring find usage in treatment of psoriasis and various other skin diseases. 2-(3,4methylenedioxyphenyl)-5-(1,3-dihydroxypropyl)-7-methoxy-benzo- furan (H) is got from the leaves of *Mechilus glaucescens* and is used in asthma, rheumatism and ulcers. Usnic acid (I) curbs growth of many Gram-positive organisms.

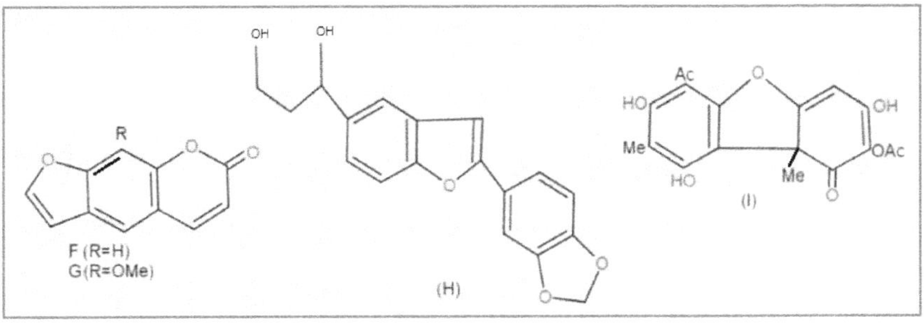

F (R=H)
G(R=OMe)
(H)
(I)

**Pterocarpans**, mentioned below, find use in antifungal activity.

## 6.6 THIOPHENE & ITS BENZO DERIVATIVES

The aromatic sextet of thiophene increases resonance stabilization due to hybridized d-orbital of sulphur.

## 6.6.1 SYNTHESIS

Iddon and Scrowston gave various methods to prepare thiophene and substituted thiophene.

The dehydration of alcohol (A) and (B) resulted in conversation to respective thiophenes.

Substituted thiophenes can be obtained from divinyl disulphides. The end products of this procedure are 3,4-di alkyl-2,5-di (methyl-thio)thiophene(a) and 5-methylthio-2- thiophene thiol(b).

**Perkin condensation** has various approaches; addition of thiogly-colic esters to 1,3- diketones or $\alpha$, $\beta$- unsaturated ketones is one such interesting approach.

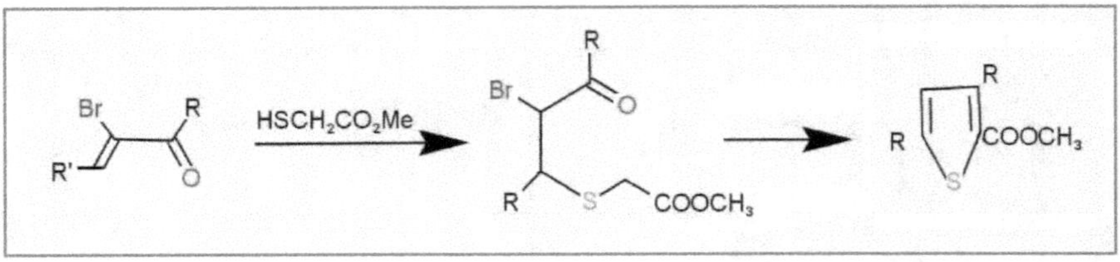

At higher temperature, an alkyl or aryl substituent on carbon atom adjacent to sulphur in propionic acid part of diester readily favours cyclization (**Dieckmann cyclization**).

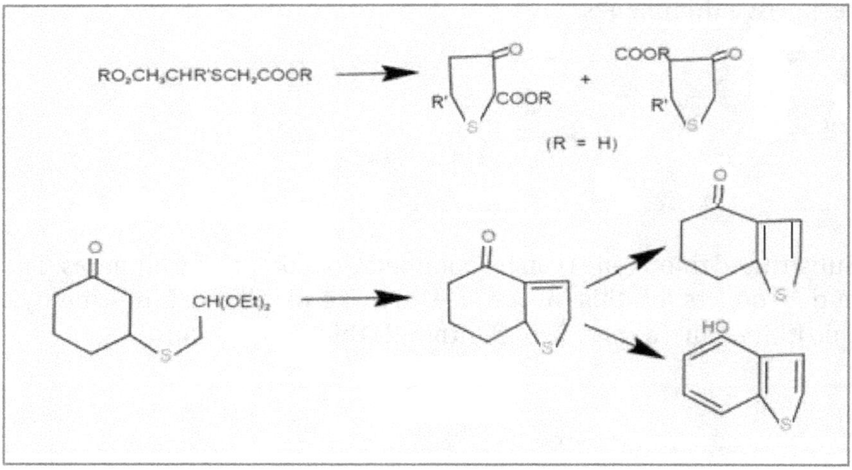

**Krubsack mechanism** explains the steps better to prepare thio-phene-2,5-di carbonyl chloride by the dehydrogenation of tetrahy-drothiophene in the presence of sulfuryl chloride. During this reac-tion sulphur dioxide and hydrogen chloride are released.

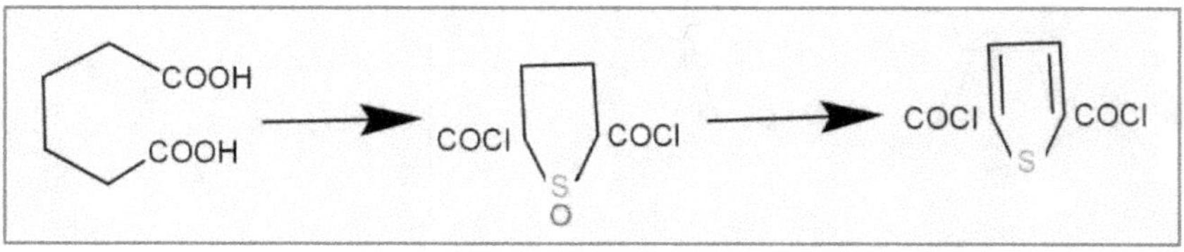

Tetra substituted thiophene can easily be obtained by cyclization reaction. In the same way, 2- aryl amino - 3 – nitro thiophene (a) is also obtained and many more such reactions have been proposed.

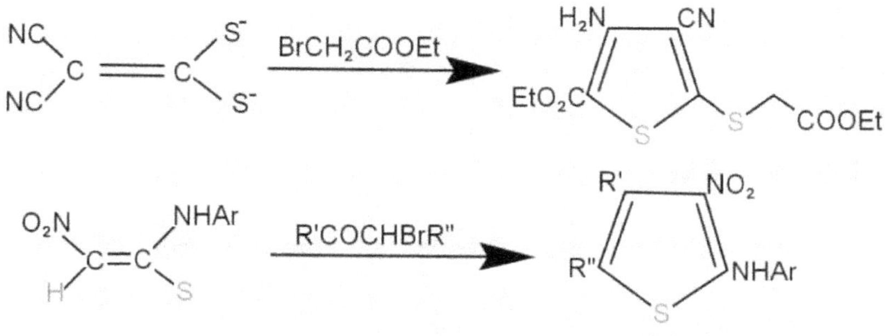

(a)

**Hinsburg ring closure** was first proposed when $\alpha$-di carbonyl compounds underwent condensation with thio glycolic esters to give substituted thiophene-2-5-dicarboxylic esters in the presence of alcoholic sodium ethoxide.

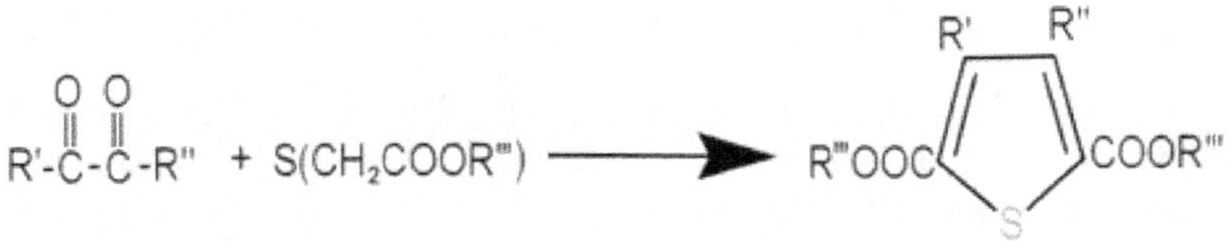

Reaction of 2- vinyl thiophene with maleic anhydride gives tetra hydro benzo thiophene which gives benzo[b]thiophene on heating with sulphur. 1,4-naphthaquinone on condensation with 3-vinylthiophene finally results in the formation of thieno anthraquinone.

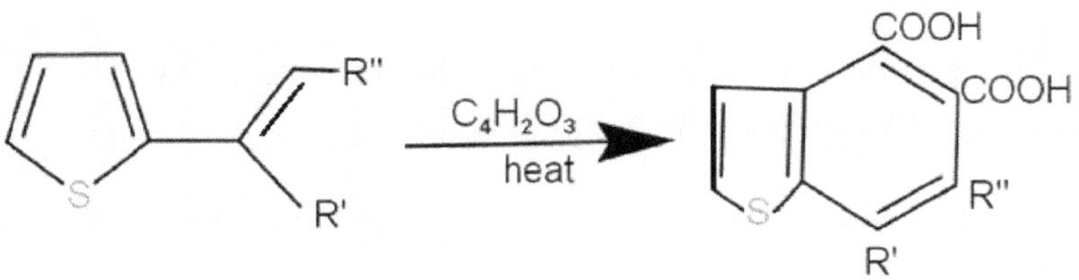

## 6.6.2 REACTIONS

Allyl thienyl sulphides via **thio-Claisen rearrangement** on being heated in quinoline when position-2 of 3-thienyl sulphide is blocked then rearrangement to position-4 occurs.

The **electrocyclization** followed by [1,5] supra-facial sigma tropic H-shift results in the pyrolysis of 1-(2-thienyl) -1-3-butadine to produce 4,5-dihydrobenzo[b]thiophene.



Tricyclic product is obtained by **thermal cyclization** of β-benzo[b]thienyl vinyl isocyanate.

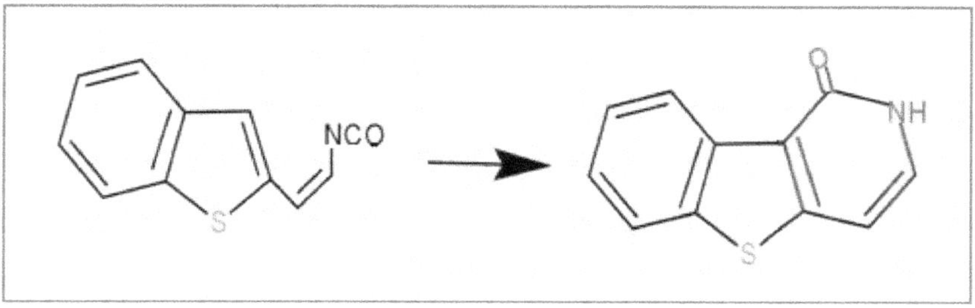

Octa hydro di benzo thiophene is obtained by **bis–cyclization** of a diene.

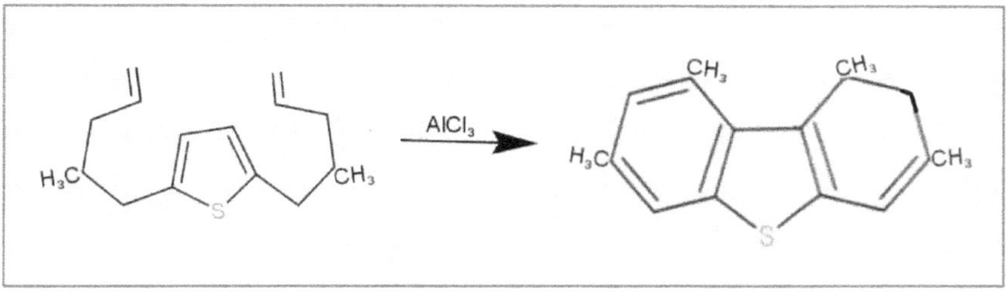

**Alkylation** of thiophene can be brought about by two routes. Under proper conditions, β-keto sulphide or an aromatized product can also be obtained from β keto sulfoxide (a).

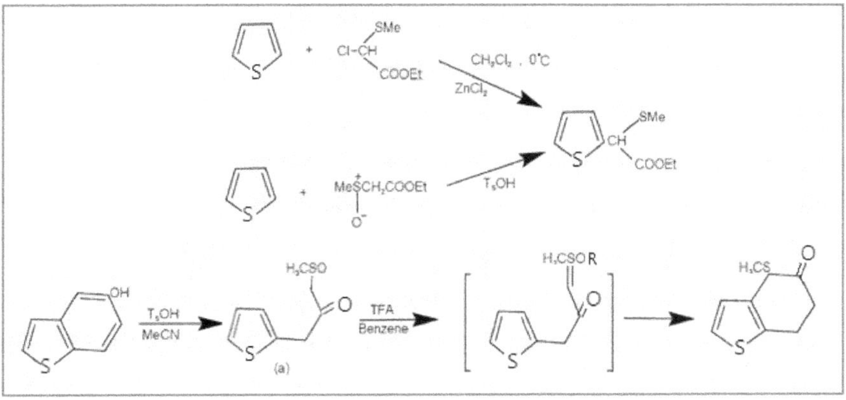

In the presence of acidic contents, thiophene tends to polymerise which may be a result of repeated **protonation** and **electrophilic substitution**.

Low yields of thieno benzo thiophenes are obtained In **Bradsher reaction**.

Fused thiophene ring systems can be generated by **intramolecular acylation**. If the side chain is attached at position-2, then ring closure take place at position-3. The reaction between 2-substituted thiophene and methacrylic acid in PPA leads to a mixture of (a) & (b). Compound (a) is formed by acylation at position-5 followed by acylative cyclization. A diacid gives dibenzothiophene derivative by double cyclization.

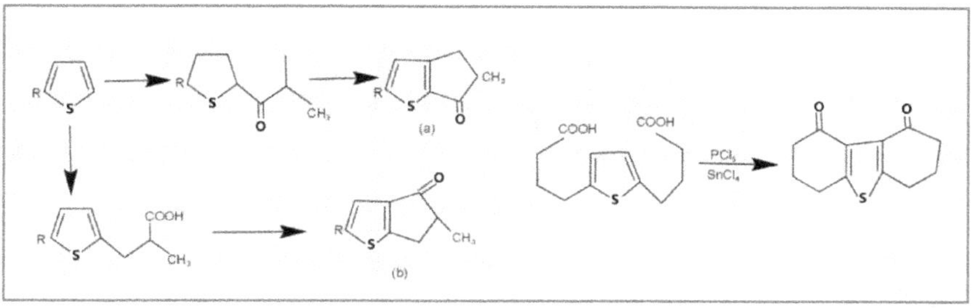

Nitration of 2-formylthiophene with $HNO_3$ and $H_2SO_4$ gives a mixture of nitro substituted compounds.

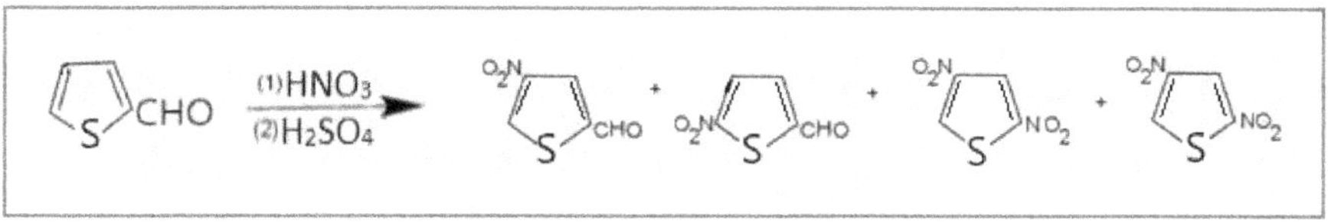

Thiophene in the presence of methanol undergoes hydrogenation and later, one of the products lead to ring opening (**Birch Reduction**).

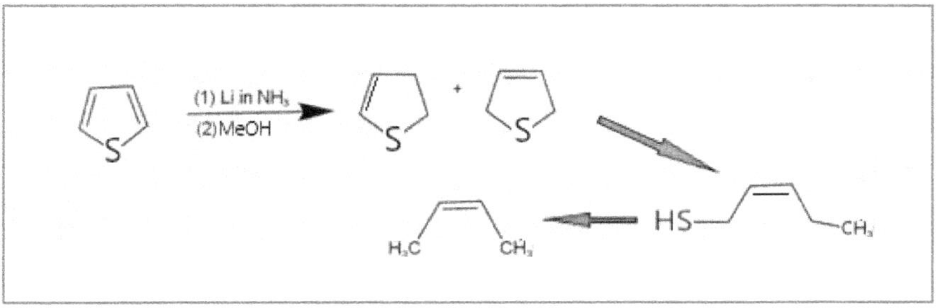

2-Substituted thiophene yields a dihydro derivative as major product along with many minor products.

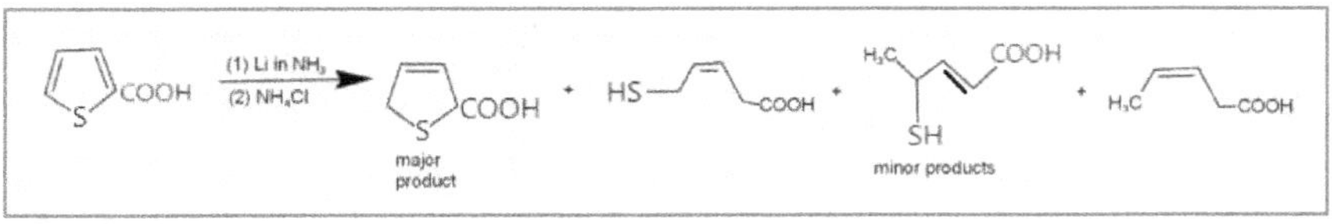

Nitrenes attack only carbon and not sulphur of thiophenes. Unsubstituted thieno pyrroles are formed by thermolysis of vinyl azides.

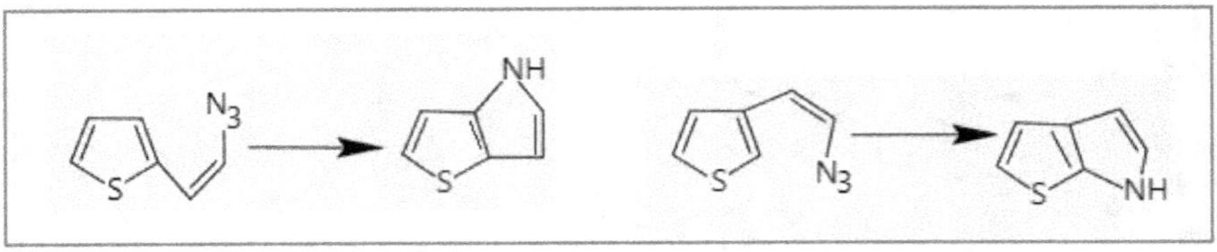

The reaction of tetramethyl thiophene with dicyano acetylene gives **[2+2] cycloadduct** in the presence of $AlCl_3$.

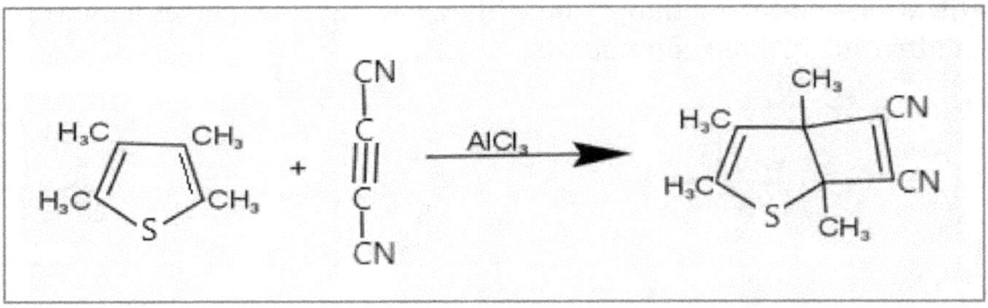

Dihalogen maleimides react with thiophenes and result in photo-substituted products.

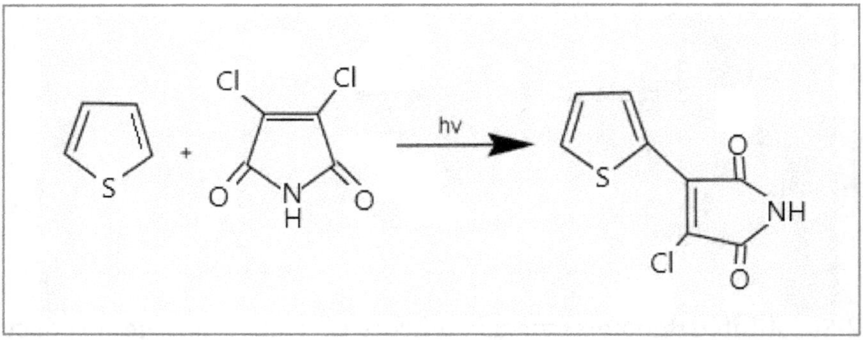

2,5 – Diamino – 3,4 – di cyano thiophene is converted to 2- amino – 3,4 – dicyano – 5 – mercapto pyrrole is the presence of base (NaOH). In this reaction, before the intermediate gets recycled, a rotation about 4,5–bond takes place.

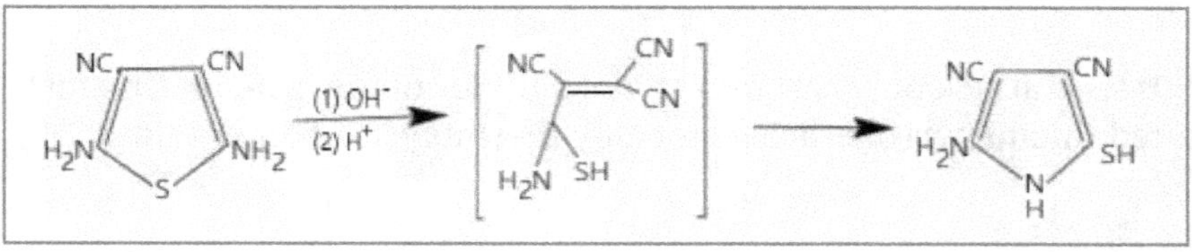

When bromo-substituted thiophene is exposed to DMSO and sodium alkoxide, partial bromination occurs whereas sodium alkoxide in DMF results in bromine migration and poly-bromination.

Lithium derivatives show a wide variety of reactivity.

**Ullman coupling** reaction of iodo-thiophenes gives tricyclic compounds in presence of copper.

## 6.6.3 APPLICATIONS

Dye industry owes a lot to the credit of thiophene and its analogues. There are various dyes of interest.

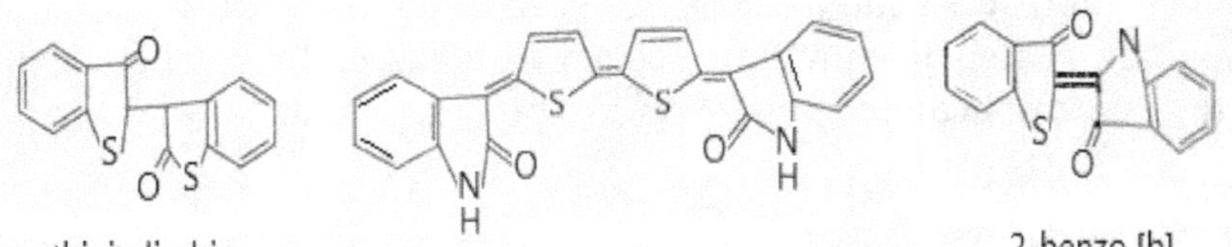

thioindirubin
(2,3' - bithioindigos)

indophenine (blue)

2-benzo [b]
thiophene-2'- indole

β- 2 – Thienyl alanine (Ar = 2–thienyl) acts as an anti-metabolite for essential amino acid.

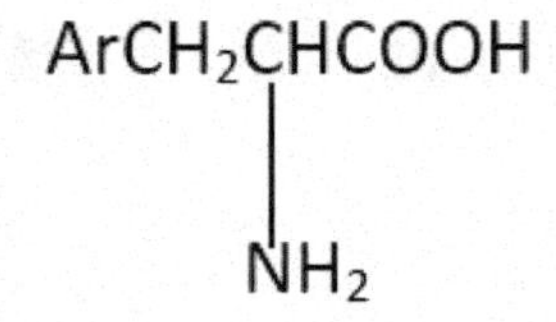

$$ArCH_2CHCOOH$$
$$|$$
$$NH_2$$

A series of aryl phenyl hydantoins (a) act as anticonvulsant whereas benzodiazepine derivative (b) commonly known as **Clozapine** acts as an agent in Central Nervous System.

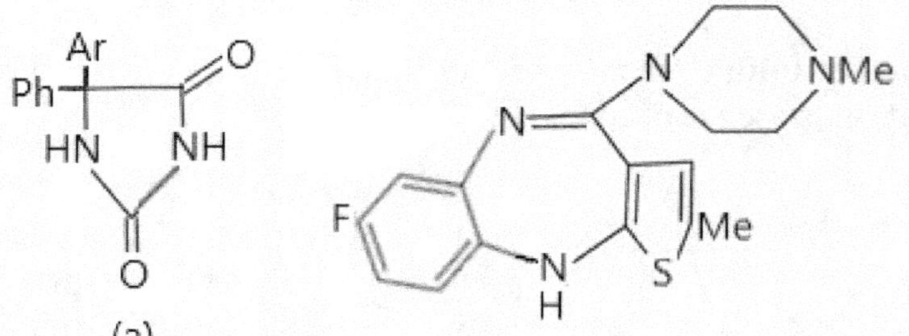

(a)

(b)

Like indole – 3 – acetic acid, **benzo (b) thiophene – 3 – acetic acid** finds use as plant growth promoter.

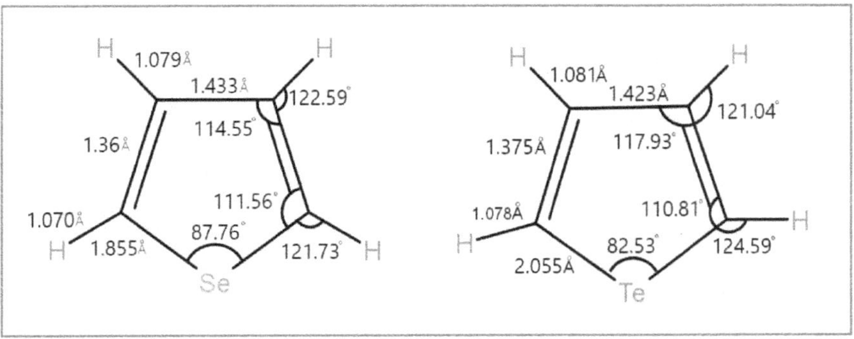

4 – N – methyl carbamoyl benzo[b]thiophene acts as an effective insecticide in the name of **Mobam**.

## 6.7 SELENOPHENES, TELLUROPHENES AND THEIR BENZO DERIVATIVES

The structure of selenophene and tellurophene finds evidence in microwave spectroscopy and H' NMR spectra.

136

## 6.7.1 SYNTHESIS

When ethyl o-methyl seleno cinnamate is treated with bromine and pyridine, we get ethyl benzoselenophene-2-carboxylate; the same method can also be used to yield 2-substituted benzo tellurophenes.

The formation of two bonds is a far more widely used approach for synthesizing selenophenes and tellurophenes.

In the presence of sodium methoxide, a mixture of 3,4-diphenylsele-nophenecarboxylic acid esters is obtained on condensation of benzil with dimethyl seleno di acetate.

## 6.7.2 REACTIONS

Selenophene responds fairly to electrophilic substitution reactions but tellurophenes give restricted reactivity due to sensitivity to acids. 2 – Thienyl selenophene (A) and 2 – benzyl selenophene (B) result in a fair yield of derivatives.

**Protonation** of both the compounds (A & B) in polyphosphoric acid give 2 - protonated products. Fuming nitric acid in acetic anhydride gives a mixture of 2-nitro and 3-nitro selenophene. α-Position of selenophene is occupied on halogenation while tri- and tetra- derivatives can also be obtained. **1,1 - Addition** products are obtained with halogens in case of tellurophenes and its derivatives. Selenophene and its derivatives give metallated products with organolithium compounds in 2-lithio derivatives. There are also variety of other products that are obtained.

Methyl substituted products are obtained when dimethyl sulphate brings about methylation of 5-methyl -2-hydroxyselenophene. Benzoselenophene-2-one also gives some good reactions.

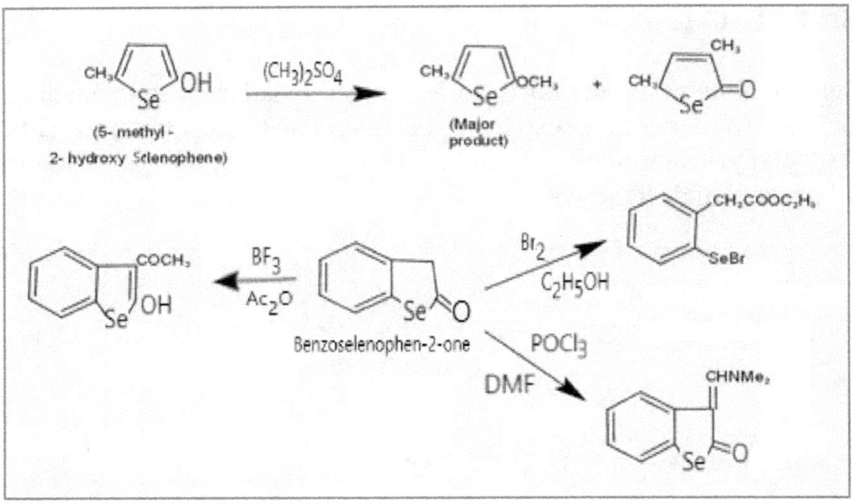

The heterocyclic ring cleavage takes place in presence of bromine in methanol.

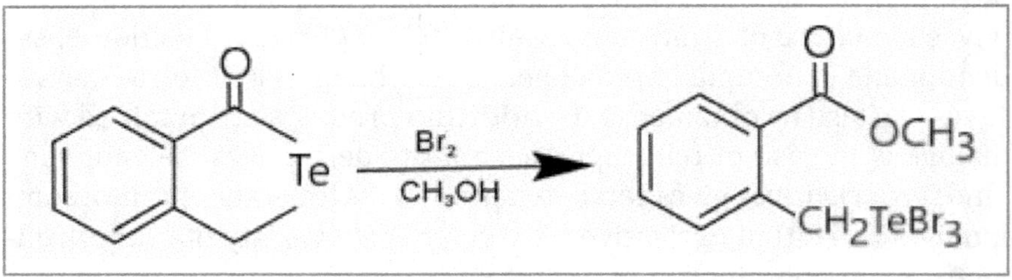

Benzotellurophen-3-one serves as starting material for the preparation of 3 - substituted derivatives.

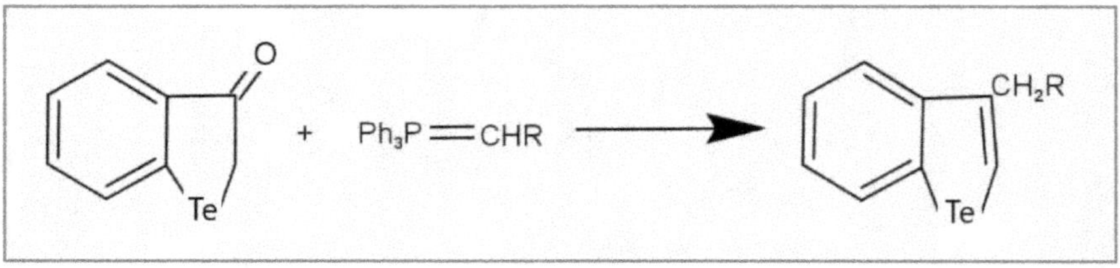

3 – bromo derivatives after lithiation results in **ring opening** products.

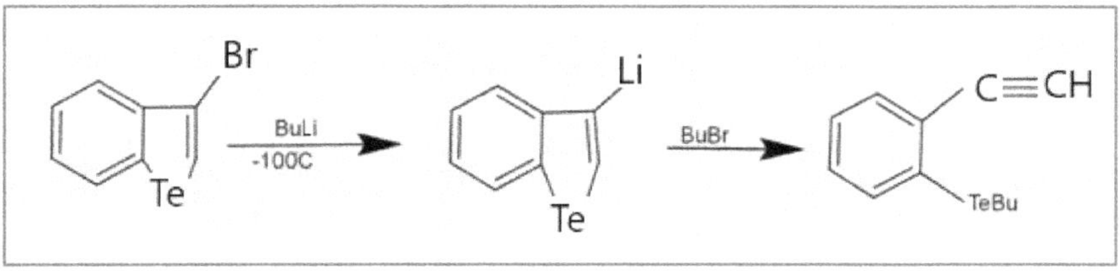

Halogenated selenophene and tellurophene with special mention to tetra choro selenophene has found use in **increasing fire resistance and autoignition temperature of aircraft hydraulic fluid**. Seleno-phene-β-ketones are found effective high temperature antioxidants for silicone fluids.

## 6.8 MISCELLENEOUS FUSED HETEROCYCLIC COMPOUNDS

### 6.8.1 INDOLIZINE

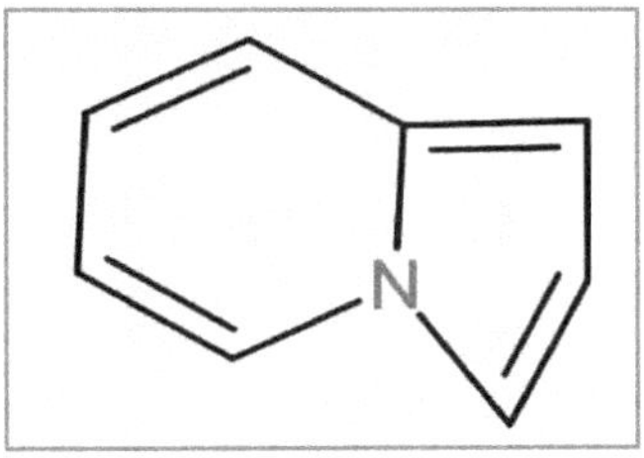

**Synthesis**

1–Phenacyl–2–picolinium bromide gets converted into a mixture of phenyl indolizine.

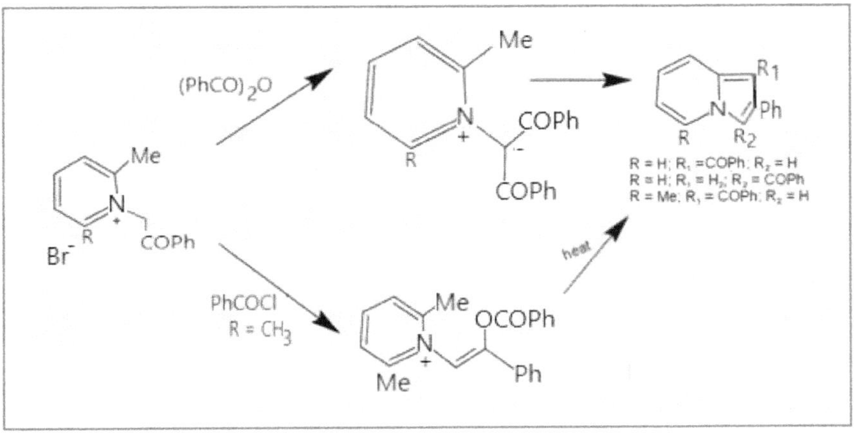

140

Substituted indolizines can be obtained by heating pyridinium salts.

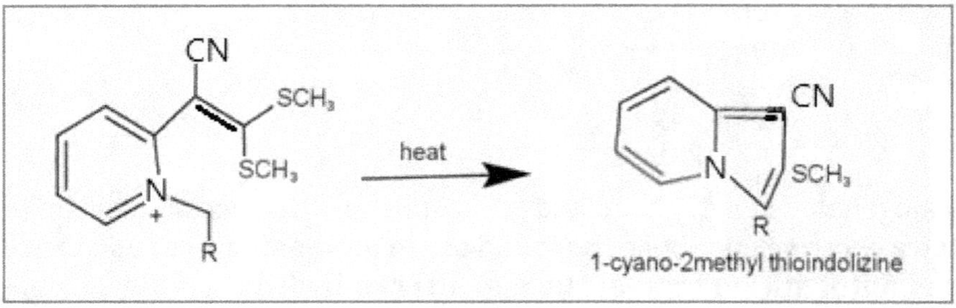

The **Scholtz synthesis** of indolizine gives a poor yield and uses acetic anhydride in the procedure.

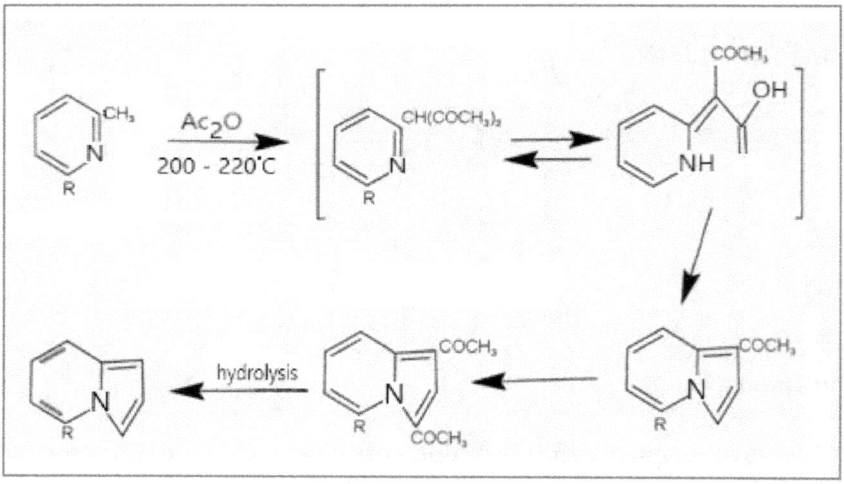

The alkaloid in *Ipomea* namely **Ipalbidine** was prepared as under.

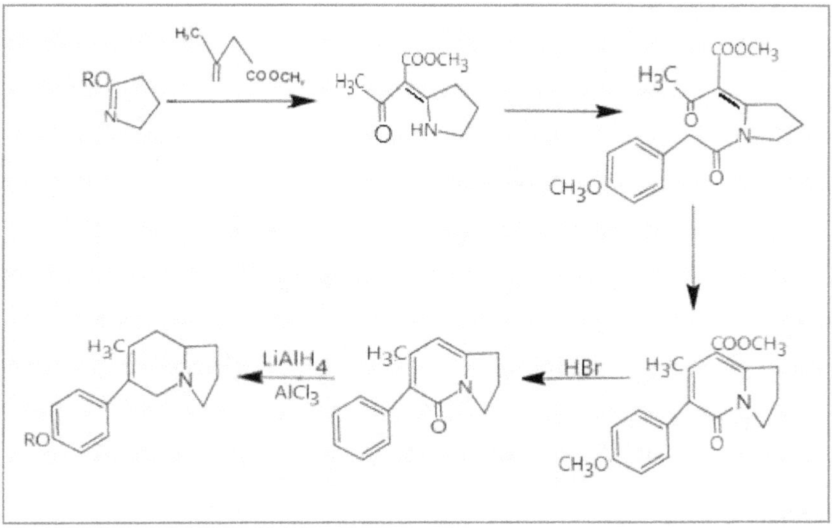

## Reactions

The nitration of 2-phenylindolizine in the presence of nitric acid and sulfuric acid gives 2-(4-nitrophenyl) indolizine (A) and 1-nitro-2-(4-nitrophenyl) indolizine (B) along with 3-nitro-2-(4-nitrophenyl) indolizine (C).

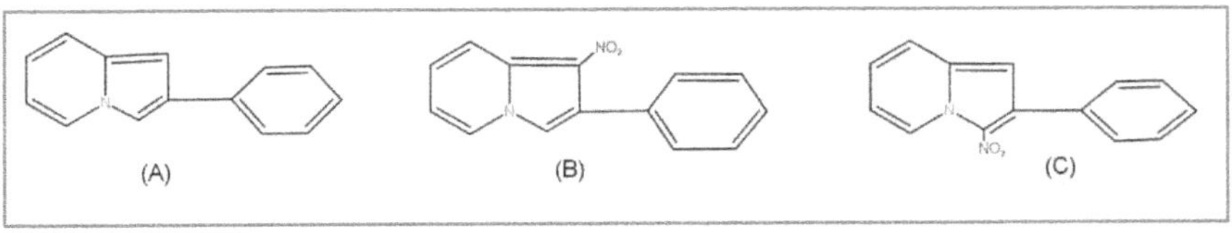

The **Mannich reaction** easily brings about substitution at position 1 and 3 in indolizine.

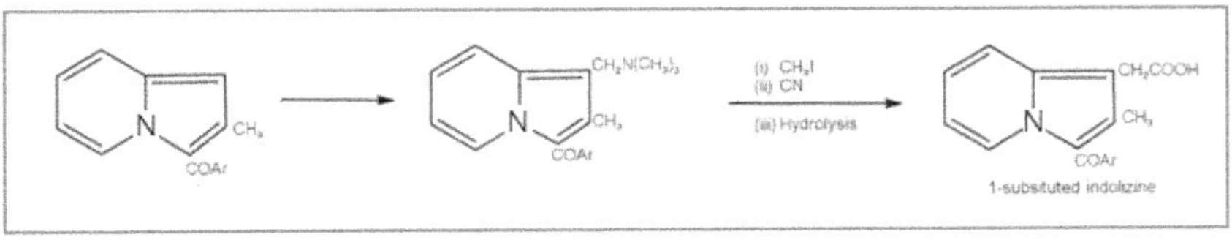

Oxidation of indolizine results in fission of ring and therefore serves as a useful means in determination of structure.

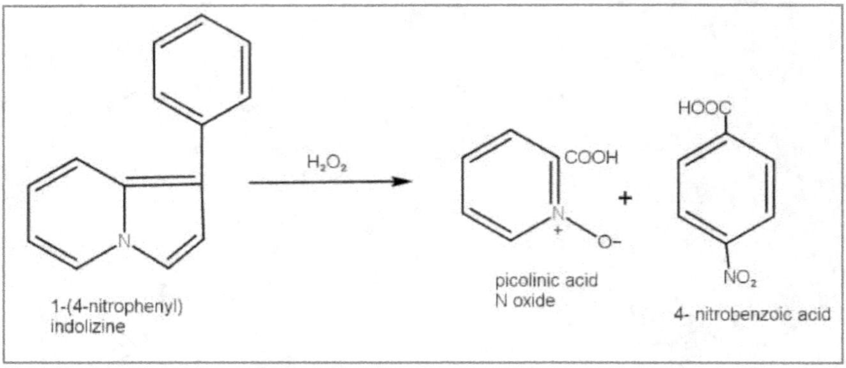

142



The reactivity of saturated indolizines have been studied and a variety of reactions of 3,5-dioxoindolizine have been put forth.

## 6.8.2 DIAZAINDENE

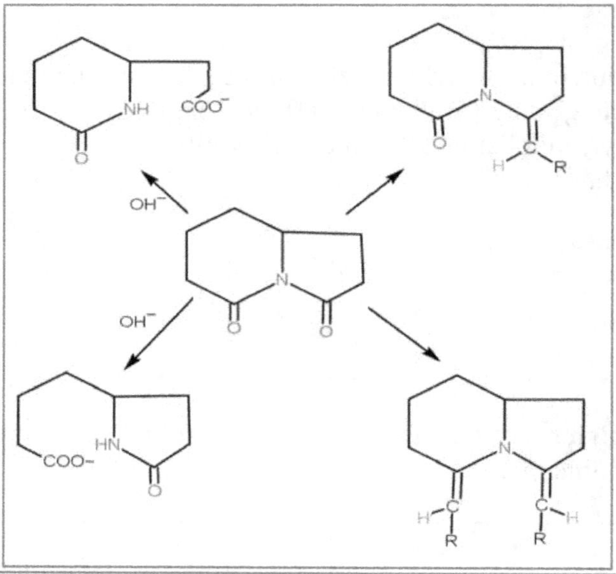

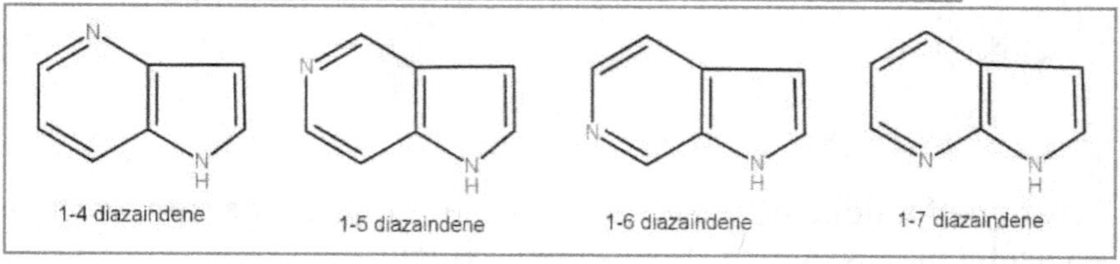

## Synthesis

The t-butyl-2-aminocyanoacetate undergoes conversion through various stages.

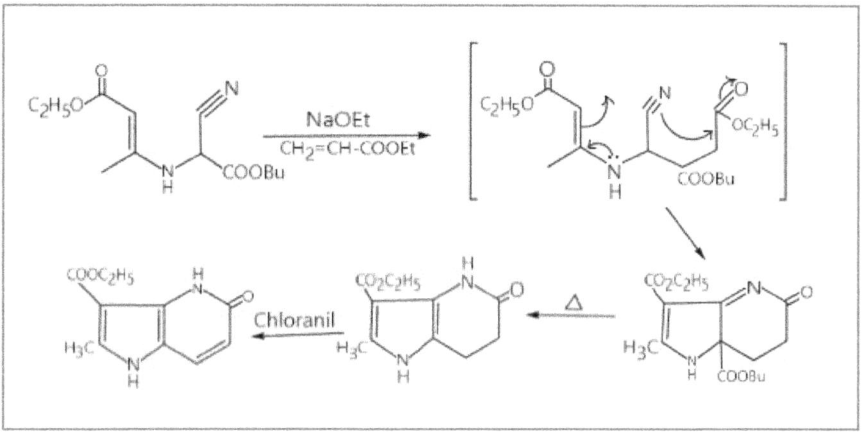

With an intact five membered heterocyclic ring, the ring closure reactions give good yield.

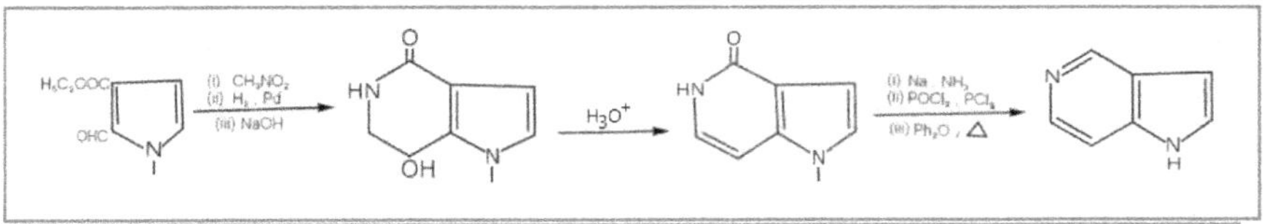

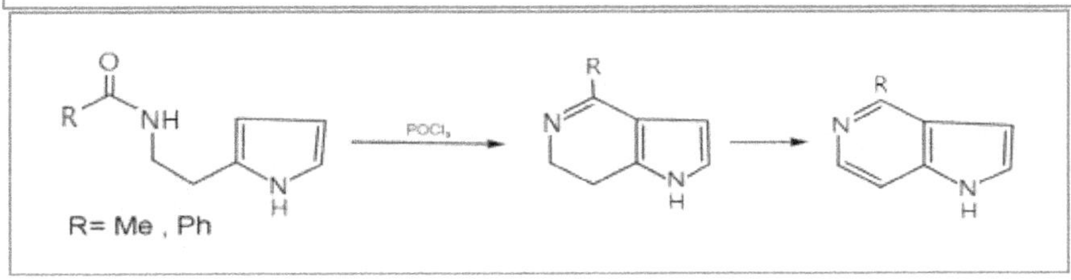

**Dim Roth Rearrangement** yield pyrrolo-pyrimidine in presence of aqueous alkali.

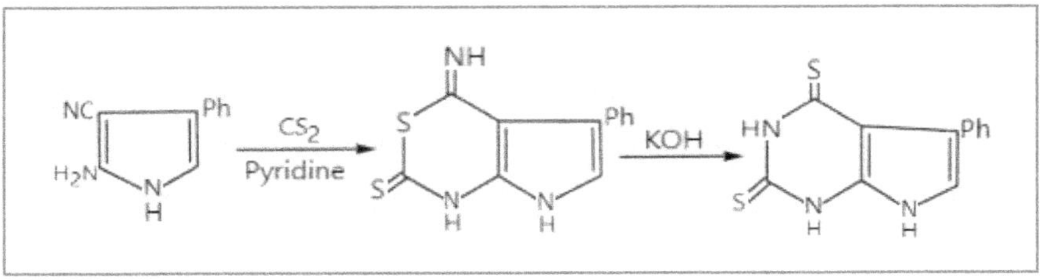

## <u>Reactions</u>

Quaternization of pyrrolo-pyridines at room temperature by methyl iodide takes place at the nitrogen of six membered ring. Further treatment with methyl iodide effects the pyrrole nitrogen also.

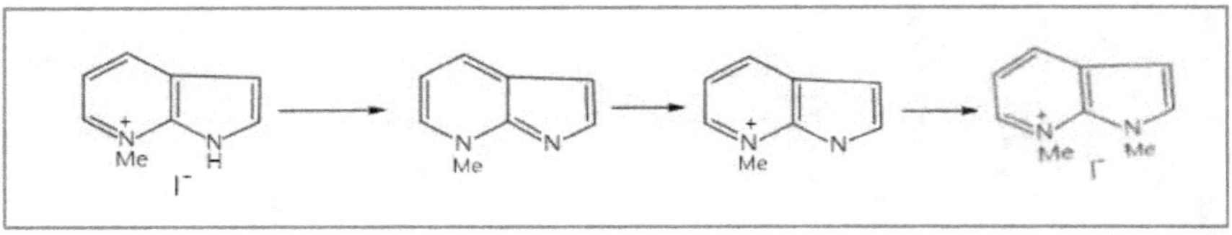

**Alkylation** results in substitution mostly at position-3 of pyrrole ring; however, mixture of 2- and 3–alkylated derivatives are also obtained.

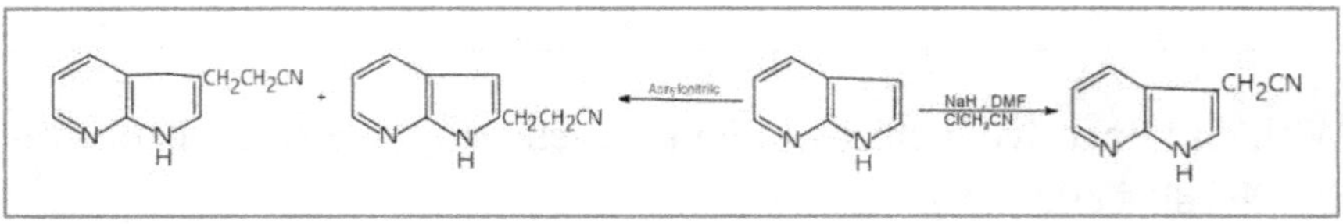

**Nitration** of 2,5-dimethylpyrrolo[3,2-b] pyridine gives 3-nitro derivatives in low quantity.

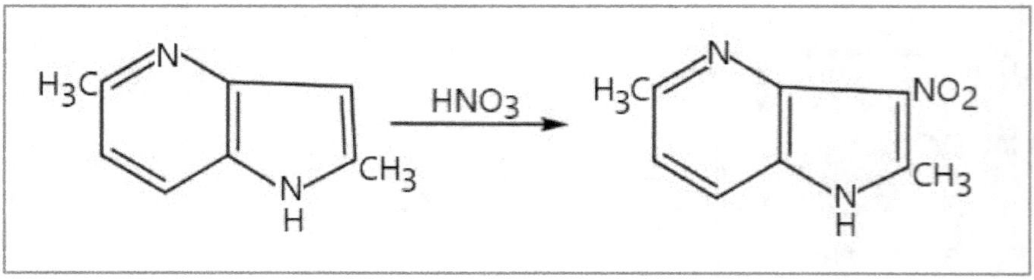

If this is a substituent at C-3, then C-2 position gives nitration product. Nitration of 3-bromopymolo[2,3-b] pyridine interestingly leads to oxidation.

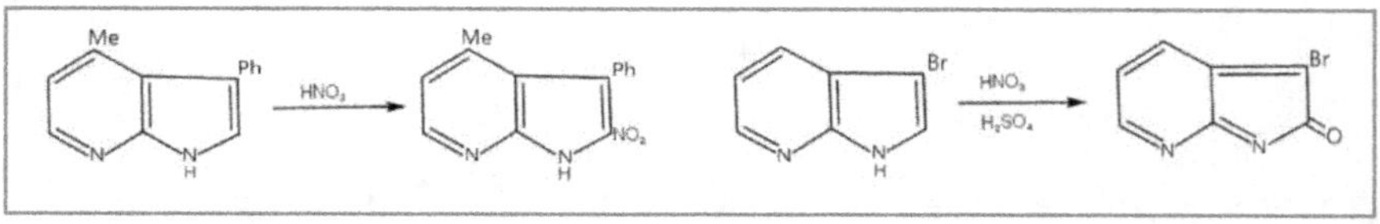

Oxidation is readily brought about in the presence of KMnO₄ however, no response is shown to silver oxide and selenium oxide.

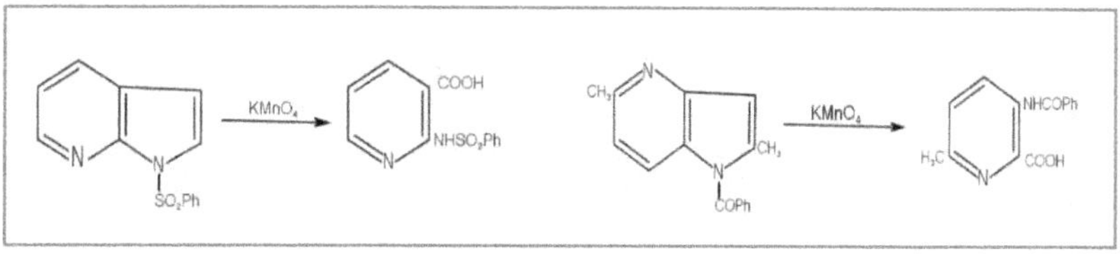

There is a series of reactions involving ring closure and rearrangement of amide resulting in pyrroloquinoline.

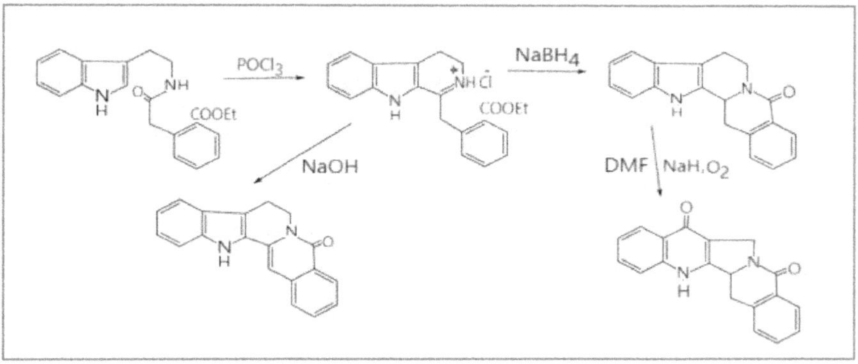

## 6.8.3 BENZOFURAN with SUBSTITUTED HETEROATOM

### Synthesis

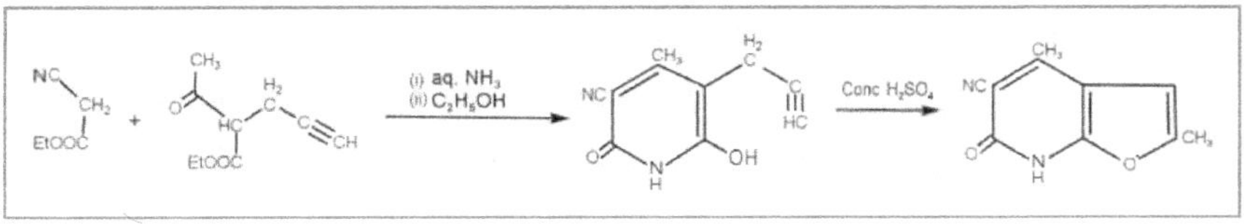

Substituted pyrinol gets converted to 1,4-dicarbonyl derivative of pyridine by methylation which on heating with concentrated hydrobromic acid leads to construction of furan ring by cyclodehydration.

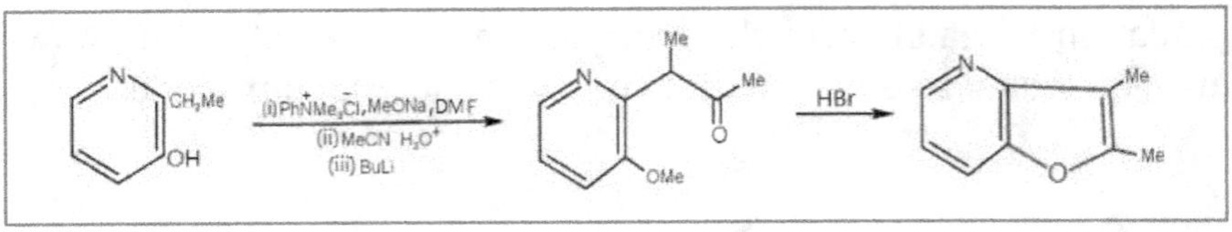

**Nucleophilic displacement** reactions take place in presence of bis (tri phenyl phosphine) nickel (II) chloride and allyl magnesium chloride.

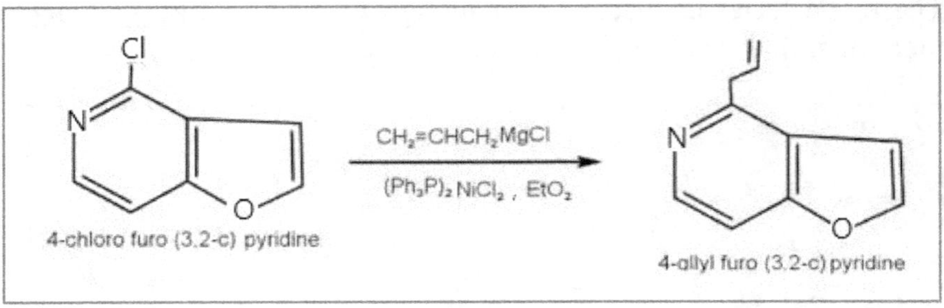

**Electrophilic substitutions** occur at furan ring.

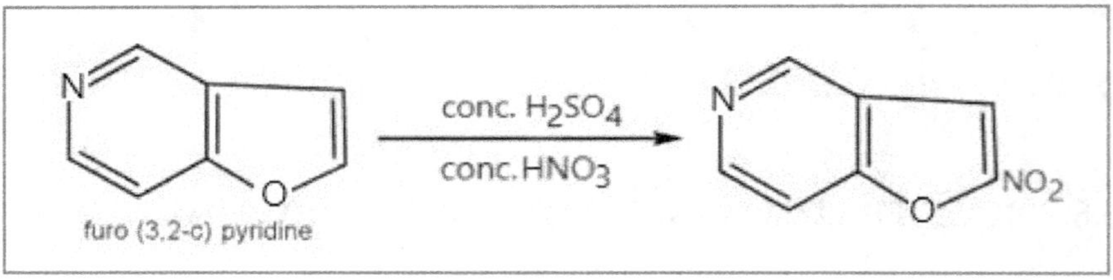

Furo pyridazine is obtained after reduction and then treatment with hydrazine or from furan-2,3-diarboxylic acid bis hydrazide.

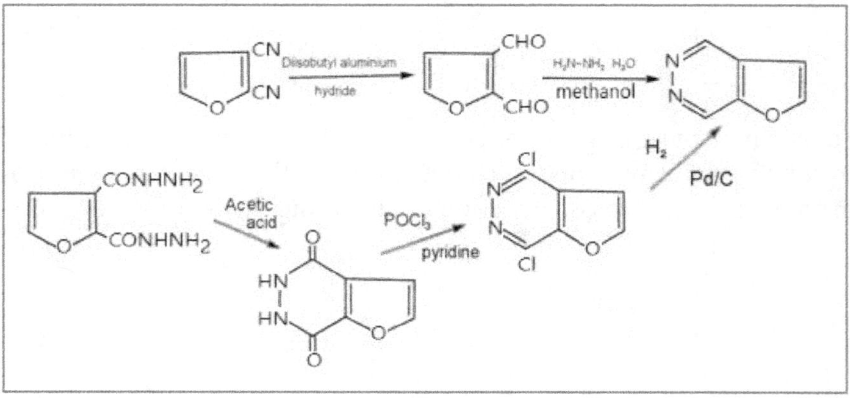

In the same way, there is another method of preparation.

Furo pyrimidine is obtained easily by ring closure reaction.

## 6.8.4 BENZOTHIOPHENE with SUBSTITUTED HETEROATOM

Substituted thienopyridine can be obtained by refluxing N-(5-methyl-2-thienyl) amino methylene malonate in $POCl_3$.

4-chloro-2-methylthieno[2,3-b]
pyridine-5-carboxylate

Pyrylium salts change to thieno[2,3-c] pyridines when treated with ethanolic ammonia. **Thermal cyclization** of 2 thienyl vinyl isocyanate results in the formation of thieno[3,2-c] pyridin-4-one. The chloro compound is then reduced to get required product. When hydrazine reacts with 2,3 – dicyano thiophene it results in the formation of 4, 7- diamino thieno (2,3-d) pyridoxine.

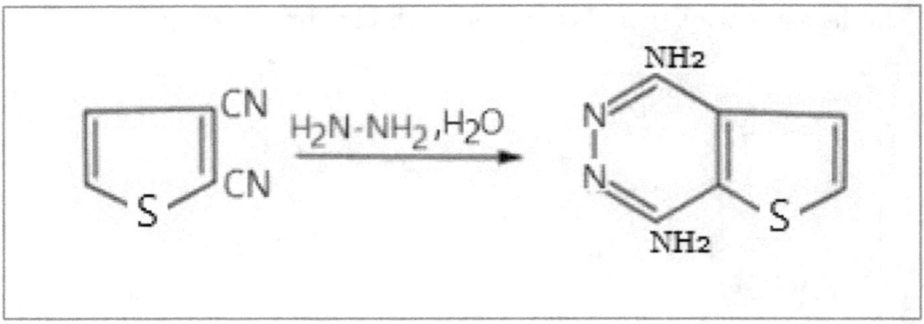

148

4-Amino thieno[2,3-d] pyrimidines can be obtained from 2-amino-3-cyano thiophenes. 2- Ethoxy methylene amino derivative acts as intermediate which on treatment with ethanolic ammonia gets converted into another intermediate, 2-formamidinonitrile that gives the final product with sodium methoxide in DMF.

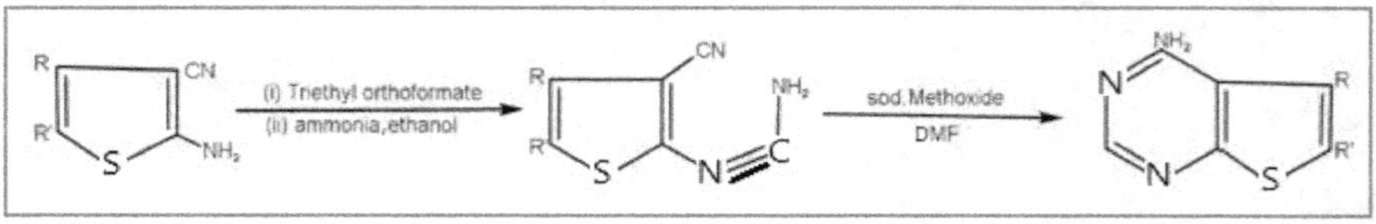

Methyl pyrazine is used as the starting material in the preparation of thieno[2,3-d] pyrazine. When this starting material is chlorinated then a mixture of 2- chloro-3-methyl pyrazine and 2-chloro-6-methyl pyrazine is obtained which finally after ring closure results in ultimate product.

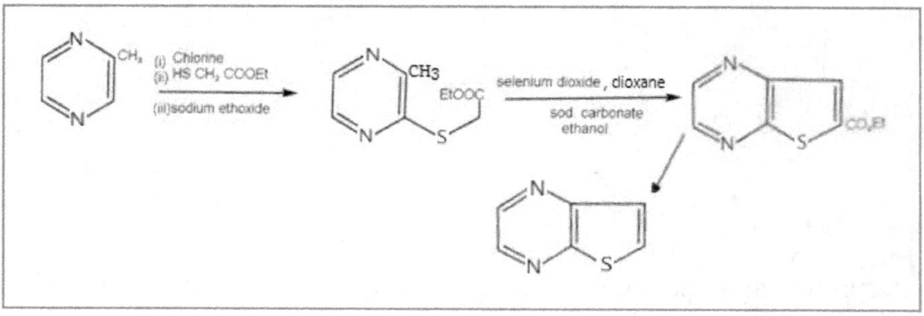

## 6.8.5 BENZOSELENOPHENE with SUBSTITUTED HETEROATOM

Preparation of ethyl-3-hydroxyselenelo[2,3-b] pyridine-2-carboxylate proceeds via several steps. The starting material for this reaction is ethyl-2- chloropyridine-3-carboxylate and methane selenol.

In another method of synthesis, cyclization followed by decarboxylation of intermediate acid results in the product. The same product can also be obtained by reduction of 2-nitro selenophene and then condensing the amino compound with malonaldehyde bis (diethyl acetal) in $ZnCl_2$.

## REFERENCES


Carra, P. O'. and Hoard, J. L. (1975); in 'Porphyrins & Metalloporphyrins', ed. K. M. Smith; Elseivier, Amsterdam, 123.
Jackson, A. H., Kenner, G. W. and Smith, K. M. (1968); J. Chem. Soc.(C), 302.
H. H. Inhoffen and W. Nolte; Tetrahedron Lett., 2185.
Bonnet, R., Dimsdale, M. J. and Stephenson, G. F. (1969); J. Chem. Soc.(C), 564.
Battersby, A. R., McDonald, E. and Hoard, J. L. (1975); in 'Porphyrins & Metalloporphyrins', ed. K. M. Smith; Elseivier, Amsterdam, 61.
Fischer, H. and Orth, H. (1937); 'Die Chemie des Pyrroles', Ahademische Verlag, Leipzig, Vol.2(1), 219.
Pozdeev, N. M., Akulinin, O. B., Shapkin, A. A. and Magdesieva, N. N. (1969); Dukl. Akad. Nauk. SSSR, 185, 384.

Brown, R. D., Burden, F. R. and Godfrey, P. D. (1968); J. Mol. Spectrosc., 25, 415.

Brown, R. D. and Crofts, J. G. (1973), Chem. Phys., 1, 217.

Dahlqvist, K-1. and Hornfeldt, A. B. (1971); Chem. Ser., 1, 125.

Annibale, A-D', Lunazzi, L., Fringuelli, F. and Taticchi, A. (1974); Mol. Phys., 27, 257.

Chidichimo, G., Lelj, F., Longeri, M., Rosso, N. and Varcini, C. A. (1980); J. Magn. Reson., 41, 35.

Marechel, G., Christiaens, L., Renson, M., Jacquignon, P. and Croisy, A. (1977); Bull. Soc. Chim. Fr., 157.

Piette, J. -L., Talbot, J. -M., Genard, J. -C. and Renson, M. (1973); Bull. Soc. Chim. Fr., 2468.

Engman, L. and Cava, M. P. (1981); J. Org. Chem., 46, 4194.


Loth - Compere, M., Luxen, A., Thibaut, Ph., Christiaens, L., Guillaume, M. and Renson, M. (1981); J. Heterocycl. Chem., 18, 343.

Backer, H. J. and Sterens, W. (1940); Recl. Trav. Chim. Pays-Bas, 59, 423.

Konstantinov, P. A., Koloskora, N. M., Shupik, R. I. and Volkov, M. N. (1973); Zh. Obsheh. Khim., 43, 872.

Yurin, Yu. K., Zaitseva, E. L. and Rozantsev, G. G. (1960); J. Gen. Chem. USSR (Engl. Transl.), 30, 2189.


Fringuelli, F., Marino, G. and Taticchi, A. (1977); Adv. Heterocycl. Chem., 21, 119.

Gronowitz, S., Johnson, I. and Hornfeldt, A.-B. – (1975); Chem. Ser., 7, 111.

Cederlund, B. and Hornfeldt, A. B. (1976); Acta Chem. Seand., Ser. B, 30, 101.

Evers, M., Weber, R., Thibaut, Ph., Christiaens, L., Renson, M., Croisy, A. and Jacquignon, P. (1976); J. Chem. Soc., Perkin Trans. 1, 2452.


Bergman, J. and Engman, L. (1979); Tetrahedron Lett., 1909.


McCord, R. S. D. H. Nail and M. B. Sheratte; Chem. Abstr., 1973, 79, 44 208.

Sheratte, M. B. (1974); Chem. Abstr., 1974, 81, 124 071.

Magdesieva, N. N. (1970); Adv. Heterocycl. Chem., 1970, 12, 1.

Litsel, W. F. and Gertsmann, E. (1972), Chem. Ber., 105, 2344.

Kakehi, A., Ito, S., Maeda, T., Takeda, R., Nishimura, M., Tamashima, M. and Yamaguchi, T. (1978); J. org. chem., 43, 4837.

Scholtz, M. (1912); Ber, 45, 734.

Wick, A. E., Bartlett, P. A and Dolphin, D. (1971); Helv. Chim. Acta, 54, 513.

Greci, L. and RIdd, J. H. (1979); J. Chem. Soc., Perkin. Trans. 2, 312.

Jones, G. (1966); In the organic chemistry of nitrogen, ed. Sidgwick, N.P., Millar I. T. and Springall, H.D.; Clarendon Press, Oxford, 3rd edn., 752.

Flitsch, W. (1964); Chem. Ber, 97, 1548.

Murata, T., Sugawara, T. and Ukawa, K. (1978); Chem. Pharm. Bull., 26, 3080.

Bisagni, E. (1972); 'The Jerusalem Symposia on Quantum Chemistry and Biochemistry', ed. Bergmann, E. D. and Pullman, B.; Academic Press, New York, Vol. 4, p. 439.

Herz, W. and Tocker, S. (1955); J. Am. Chem. Soc., 1955, 77, 6355.

Bell, M. R. (1965); Belg. Pat. 659 467 (Chem. Abstr., 1966, 64, 2098).

Herbert, R. and Wibberley, D. G. (1969); J. Chem. Soc. (C), 1505.

Kruber, O. (1943); Ber., 76, 128.

Clayton, J. C. and Kenyon, J. (1950); J. Chem. Soc., 105, 2952.

Rodionov, V. (1926); Bull. Soc. Chim. Fr., 39, 305.

Baker B. R. and McEroy, F. J. (1955); J. Org. Chem., 20, 136.

Hymans, W. E. and Cruickshank, P. A. (1974); J. Heterocycl. Chem., 11, 231.

Knabe, J. and Heckmann, R. (1980); Arch. Pharm. (Wein-heim, Ger.), 313, 1048.

Knabe, J. and Heckmann, R. (1981); Arch. Pharm. (Wein-heim, Ger.), 314, 156.

Mcfarland, K. W., Essary, W. E., Cilenti, L., Cozart, W. and Mcforland, P. E. (1975); J. Heterocycl. Chem., 12, 705.

Bleackley, R. C., Jones, A. S. and Walker, R. T. (1976); Tetrahedron, 32, 2795.

Gillis, P. M., Haemers, A. and Bollaert, W. (1980); Eur. J. Med. Chem.–Chim. Ther., 15, 185.

Dulenko, L. V., Dorofcenko, G. N., Baranov, S. N., katts, I. G. and Du-lenko V. I. (1971); Khim. Geterotsiki. Socdin., 320 (Chem. Abstr. 1972, 76, 14 370.

Eloy, F. and Deryckere, A. M. (1970); Bull. Soc. Chim. Belg., 79, 301.

Quturquin, F., Ah-kow, G. and Paulmier, C. (1973); C. R. Hebd. Seances, Acad. Sci. Ser. C, 277, 29.

Quturquin, F., Ah-kow, G. and Paulmier, C. (1976), Bull. Soc. Chim. Fr.,88.

Taylor, E. C. and Berger, J. E. (1967); J. Org. Chem., 32, 2376,

Bourguignon, J., Lemarchand, M. and Queguiner, G. (1980); J. Heterocycl. Chem., 17, 257.

Pirson, P. and Christiaens, L. (1973); Bull. Soc. Chim. Fr., 704.

Quturquin, F., Ah-kow, G. and Paulmier, C. (1976); Bull. Soc. Chim. Fr., 883.

Benary, E. and Konrad, R. (1923); Ber., 56, 44.

Sugasawa,T., Adachi,M., Sasakura,K. and Kilagawa,A. (1979); J. Org. Chem., 44, 578.

Smith, A. B., Levenberg, P. A., Jerris, P. J., Scarborough, R. M. Jr. and Wovkulich, P.M. (1981); J. Am. Chem. Soc., 103, 1501.

Capuano, L. and Fischer, W. (1976); chem. Ber., 109, 212.

Hamaguchi, M. and Ibata, T. (1976); Chem. Lett., 287.

Miller, J. A. and Woods, H. C. S. (1968); J. Chem, Soc. (C), 1837.

Brust, D. P., Tarbell, D. S., Hecht, S. M., Hayward, E. C. and Colebrook, L. D. (1966); J. Org. Chem., 32, 2192.

Kwart, H. and Euans, E. R. (1966); J. Org. Chem, 31, 413.

Kataev, E. G., chmutora, G. A., Musina, A. A. and Anstas' eva, A. P. (1967); Zh. Org. Khim., 3, 597.

Anderson, W. K., LaVoie, E. J. and Bottaro, J. C. (1976); J. Chem. Soc; Perkin Trans. 1, 1.

Marechal, G., Christiaens, L., Renson, M., Jacquignon, P. and Croisy, A. (1977); Bull. Soc. Chim. Fr., 157.

Severin, T., Adhikary, P. and Schnabel, I. (1969); Chem. Ber., 102, 1325.

Baxter, I. and Swan, G. A. (1968); J. Chem. Soc.(C), 468.

Mulligan, P. J. and LaBerge, S. (1970); J. Med. Chem, 1970, 13, 1248.

Pummerer, R. and Marondel, G. (1956); Chem. Ber., 89, 1454.

Backett, A. H., Linstead, R. P. and Walker, J. (1968); Tetrahedron, 24, 6093.

Barrett, P. A., Linstead, R. P. and Tuey, G. A. P. (1939); J. Chem. Soc., 1809.

Yokoyama, M., Kurauchi, M. and Imamoti, T. (1939) Tetrahedron Lett., 1809.

Gewald, K. and Jansch, H. J. (1973); J. Prakt. Chem., 315, 779.

Schroeder, D. C., Corcoran, P. O., Holden, C. A. and Mulligen, M. C. (1962); J. Org. Chem., 27, 586.

Elix, J. A. and Murphy, D. P. (1975); Aust. J. Chem., 28, 1559.

Zeller, K. P. and Petersen, H. (1975); Synthesis, 532.

Carruthers, W. (1966); Chem. Commun., 272.

Chakraborty, D. P. (1966); Tetrahedron Lett., 661.

Akermark, B., Eberson, L., Jonsson, E. and Pettersson, E. (1975); J. Org. Chem., 40, 1365.

Popp, F. D. (1975); Adv. Heterocycl. Chem., 18, 337.

Miller, B. and Matjeka, E. R. (1980); J. Am. Chem. Soc., 102, 4772.

Mann, M. E. and White, J. D. (1973); Chem. Commun., 5185.

Emmett, J. C., Veber, D. F. and Lwowski, W. (1965); Chem. Commun., 272.

Cairns, T. L., Carboni, R. A., Coffman, D. D., Engelhardt, V. A., Heckert, R. E., Little, E. L., McGeer, E. G., Mckusick, B. C., Middleton, W. J., Scribner, R. M., Theobald, C. W. and Winberg, H. E. (1958); J. Am. Chem. Soc., 80, 2775.

Nasakin, O. E., Alekseev, V. V., Promonenkorv, V. K., Abramov, I. A. and Bulai, A. kh. (1981); Zh. Org. Khim., 17, 1958.

Cotterill, W. D., France, C. J., Livingstone, R., Atkinson, J. R. and Cottam, J. (1992); J. Chem. Soc., Perkin Trans. 1, 192, 787.

Leonard, N. J. and Figueras, J. (1952); Jr., J. Am. Chem. Soc., 74, 917.

Braun, J. V. and Wessbach, K. (1929) Ber., 62, 2416.

Hertog, H. J. den and Buurman, D. J. (1972);Recl. Trav. Chim. Pays-Bas, 91, 841.

Sanders, G. M., Van Dijk, M. and Hertog, H. J. den (1974); Recl. Trav. Chim. Pays-Bas, 93, 298.

Peters, F. and Simonis, H. (1908); Ber., 41, 830.

Zagorevskii, V. A., Savel'ev, V. l., Meshcheryakova, L. M. and Vinokuorv, V. G. (1970); Khim. Geterotsikl. Soedin. Sb., no.2, 166 (Chem. Abstr., 1972, 76, 140 378)

Gross, G. and Wentrup, C. (1982); J. Chem. Soc., Chem. Commun., 360.

Ban,Y., Yoshida, K., Goto, J. and Oishi, T. (1981); J. Am. Chem. Soc., 103, 6990.

Warrener, R. N., Pitt, I. G. and Russel, R. A. (1982); J. Chem. Soc., Chem. Commmun., 1195.

Belen'kii, L. I., Yakubov, A. P. and Gol'dfarb, Ya. L. (1975); Zh. Org. Khim., 11, 424.

Loader, E. E. and Anderson, H. J. (1981); Can. J. Chem., 59, 2673.

Lohaus, G. (1970); Org. Synth., 1970, 50, 52.

Sundberg, R. J. (1970); 'The Chemistry of Indoles'; Academic Press, New York, 1970, pp. 182 - 183.

Blenderman, W. G., Joullie, M. M. and Preti, G. (1979); Tetrahedron Lett., 4985.

Kumar, G., Rajagopalan, K., Swaminathan, S. and Balasubramanian, K. K. (1981); Indian J. chem., Sect. B, 20, 271.

154


Reinhoudt, D. N., Volger, H. C., Kouwenhoven, C. G., Wynberg, H. and Helder, R. (1972); Tetrahedron Lett., 5269.

Wamhoff, H. and Hupe, H. J. (1978); Tetrahedron Lett., 125.

Middleton, W. J., Engelhardt, V. A. and Fisher, B. S. (1958); J. Am. Chem. Soc., 80, 2822.

Barker, J. M., Coutts, I. G. C. and Huddleston, P. R. (1972); J. Chem. Soc, Chem. Commun., 615.

Gschwend, H. W. and Rodriguez, H. R.  Org. React., 1979, 26, 1.

Brown; R. K. (1972); Chem. Heterocycle. Compd., 1972, 25-1, 479.

Harbuck, J. W. and Rapoport, H. (1971); J. Org. Chem., 36, 853.

Allen, G. R. Jr., (1973); Org. React., 20, 337.

Padwa, A., Dean, D. and Qine, T. (1975); J. Am. Chem. Soc., 97, 2822.

Jones, R. A., Marriott, M. T. P, Rosenthal, W.P. and Arques, J. S. (1980); J. Org. Chem., 45, 4514.

Jones, R. A. and Arques, J. S. (1981); Tetrahedron, 37,1597.

Jones, R. A. and Bean, G. P. (1977); The Chemistry of Pyrroles, Academic Press, London, 1977, Chap. 2.

Rush, K. (1972); Chem. Heterocycl. Compd., 1972, 25-2, 537.

Erikson, K. L., Brennan, M. R. and Namnum, P. A. (1981); Synth. Commun., 11, 253.

Ishizumi, K., Shiori, T. and Yamada, S. (1967); Chem. Pharm. Bull., 15, 863.

Remers, W. A. (1979); Chem. Heterocycl. Compd., 25-3, 357.

Jones, R. A. and Bean, G. P. (1977); The Chemistry of Pyrroles, Academic Press, London, Chap. 4.

Remers, W. A. (1972); Chem. Heterocycl. Compd., 25-1, 1.

Botta, M., Angelis, F. De and Nicoletti, R. (1979); J. Heterocycl. Chem., 16, 501.

Zee, S. H. and Chen, C. S. (1974); J. Chin. Chem. Soc. (Taipei), 21, 229 (Chem. Abstr., 1975, 82,125 215)

Gossauer, A. (1974); "Die Chemie der Pyrrole", Springer Verlag, Berlin, 1974, chap. 3.

Brogden, R. N., Heel, R. C., Speight, T. M. and Avery, G. S. (1978) Drugs, 15, 429.

Gillet, C., Hehoux, E., Kestens, J., Roba, J. and Lambelin, G. (1976); Eur. J. Med. Chem., Chim. Ther., 11, 173.

Arima, K., Imanska, H., Sosaka, M., Fukuta, A. and Tamura, G. (1964); Agric. Biol. Chem., 28, 575.

Crooke, S. T. and Bradner, W. T. (1976); Cancer Treatment Rev., 3, 121.

Mata, F., Martin, M. C. and Serensen, G. O. (1978); J. Mol. Struct., 48, 157.

Larsson, F. C. V., Brandsma, L. and Lawesson, S. O. (1974); Recl. Trav. Chim. Pay-Bas, 93, 258.

Fiesselmann, H. (1960); Angew. Chem., 72, 573.

Martani, A. (1959); Ric. Sci., 29, 520 (Chem. Abstr. 1960, 54, 4535)

Napier, R. P., Kaufman, H. A., Driscoll, P. R., Glick, L. A., Chu, C. C. and Foster, H. M. (1970); J. Hetercycl. Chem., 7, 393.

Higa, T. and Krubsack, J. (1975); J. Org. Chem., 40, 3037.

Oka, K. (1979); Heterocycles, 12, 461.

Gompper, R., Kutter, E. and Topfl, W. (1962); Leibigs Ann. Chem., 659, 90.

Merchant, J. R. and Shankarnarayan, V. (1979); Curr. Sci., 48, 585 (Chem. Abstr., 1979,91, 140 696).

Hartough, H. D. (1952); Chem. Heterocycl. Comp., 3, 1.

Tominaga, Y., Lee, M. L. and Castle, R. N. (1981); J. Heterocycl. Chem., 18, 967.

Mortensen, J. Z., Hedegaard, B. and Lowesson, S. -O. (1971); Tetrachedron, 27, 3831.

Rosen, B. L and Weber, W. P. (1977); Tetrahedron Lett., 151.

Eloy, F. and Deryckere, A. (1970); J. Heterocycl. Chem., 7, 1191.

Gourier, J. and Canonne, P. (1970); Can. J. Chem., 48, 2587.

Tamura, Y., Choi, H. -D., Shindo, H., Uenishi, J. and Ishibashi, H. (1981); Tetrahedron Lett., 22, 473.

Oikawa, Y., Setoyama, O. and Yonemitsu, O. (1974); Heterocycles, 2, 21.

Sone, T., Shiromaru, O., Igarashi, S., Kato, E. and Sawara, M. (1979); Bull. Chem. Soc. Jpn., 52, 1126.

Ahmed, M., Ashby, J. and Meth-Cohn, O. (1970); Chem. Commun., 1094.

Gol 'dfarb, Ya. L., Taits, S. J. and Belen'Kii, L. I. (1963); Tetrahedron, 19, 1851.

Frejd, T. and Karlsson, O. (1979); Tetrahedron, 35, 2155.

Rahman, A. V. and Tombesi, O. L. (1968); Tetrahedron Lett., 3925.

Chaichit, N. and Gatchouse, B. M. (1981); Cryst. Struct. Commun., 10, 83.

Krasnoslobodskaya, L. D. and Gol'dfarb, Ya. L. (1969); Russ. Chem. Rev. (Engl. Transl.), 38, 389.

Kretchmer, R. A. and Laitar, R. A. (1978); J. Org. Chem., 43, 4596.

Sargent, M. V. and Stransky, P. O. (1982); J. Chem. Soc., Perkin Trans. 1, 1605.

Boyd, D. R. and Ladhams, D. E. (1928); J. Chem. Soc., 2089.

Steinkopf, W. and Hanske, W. (1939); Liebigs Ann. Chem., 541, 238.

Garcia Gonzalez, F. (1956); Adv. Carbohydrate Chem., 11, 97.

Robinson, G. C. (1966); J. Org. Chem., 31, 4252.

Rahn, C. H., Sand, D. M., Wedmid, Y., Schlenk, H., Krick, T. P. and Glass, R. L. (1979) J. Org. Chem., 44, 3420.

Davies, J. S. H. and Deegan, T. (1950); J. Chem. Soc., 3202.

Davies, J. S. H., McCrea, P. A., Norris, W. L. and Ramage, G. R. (1950); J. Chem. Soc.,3206.

Janda, M., Srogl, J., Stibor, I., Nemee, M. and Vopatrna, P. (1973); Tetrahedron, 637.

Buchi, G., Demole, E. and Thomas, A. F. (1973); J. Org. Chem., 38, 123.

Wong, H., Chapuis, J. and Monkovic, P. (1974); J. Org. Chem., 39, 1042.

Bender, D. R., Hearst, J. E. and Rapoport, M. (1979); J. Org. Chem., 44, 2176.

Nore, P. and Honkanen, E. (1980); J. Heterocycl. Chem., 17, 985.

Stark, J. B., Walter, E. D. and Owens, H. S. (1950); J. Am. Chem. Soc., 72, 1995.

Iddon, B. and Scrowston, R. M. (1970); Adv. Heterocycl. Chem., 11, 177.

Campaigne, E. and Abe, Y. (1975); J. Heterocycl. Chem., 12, 889.

Pailer, M., Grunhaus, H. and Stof, S. (1976); Monatsh. Chem., 107, 521.

White, R. L. Jr. (1967); Ph. D Thesis, Indiana Univ.

Fletcher, T. L., Pan, H. L. Cole, C. A. and Namkung, M. J. (1974); J. Heterocycl. Chem., 11, 815.

# CHAPTER 7

# 5-MEMBERED RING WITH MULTIPLE HETEROATOMS

## 7.1 INTRODUCTION

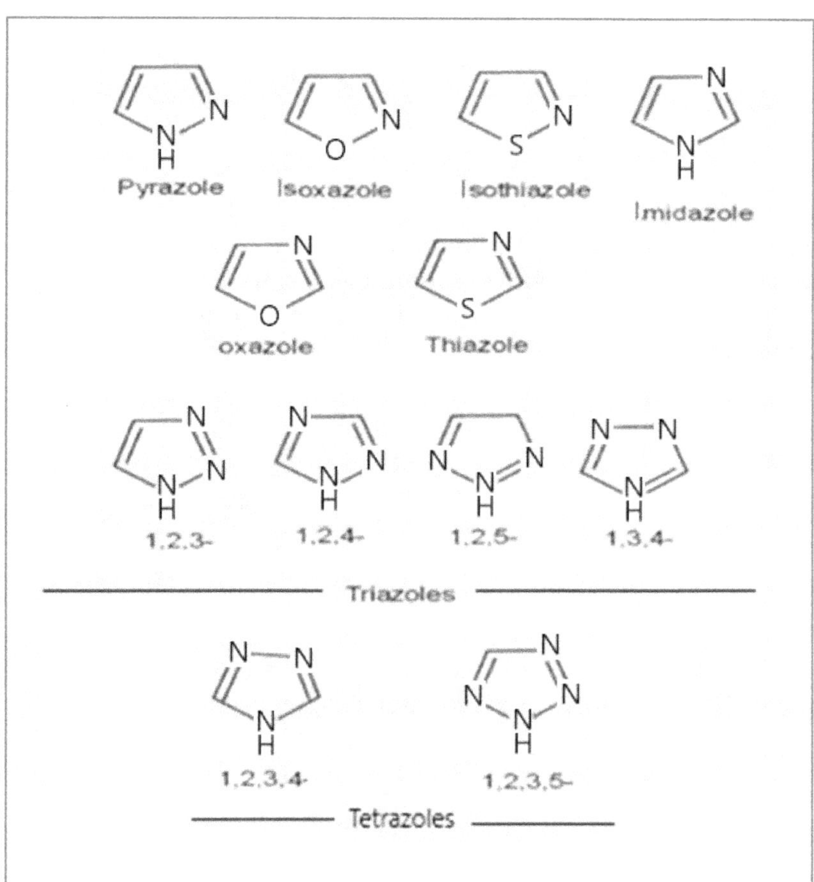



## 7.2 GENERAL METHODS OF SYNTHESIS

The general methods of synthesis of five membered rings containing two or more heteroatoms have focussed on reactions involved in forming the bonds in making up the heterocycle by **Aldol – type condensations**.

158

**Robinson - Gabriel synthesis** (as it is known collectively) results in the formation of 5-ethoxy-4-methoxyoxazole (C) from ethyl-2-formamidopropionate (B), which is prepared from alanine (A) by formylation and esterification.

2,4,5-Triarylimidazoles (E) can be prepared from α-acylamino-ketones (D) on being heated with ammonium acetate.

Three heteroatoms in a ring can be incorporated by ring closure of 1,4-dicarbonyls. 2,5-Disubstituted 1,3,4-oxadiaoles (F) are prepared by this method.

159

When bromo-pyruvaldehyde oxime (G) reacts with thiourea in methanol at room temperature, then after neutralization with sodium carbonate 2-aminothiazole-4-carbaldehyde oxime (H) is obtained.

**Cyclo-condensation** of methylhydrazine and 4,4-dimethoxybutan-2-one (I) results in the production of 1,3-dimethylpyrazole (J) and 1,5-dimethylpyrazole (K).

When methylhydrazone of acetophenone reacts with dichloromethyleneiminium salts, 5-dimethyl aminopyrazoles (L) are obtained. However, with butyl lithium, pyrazole is obtained.

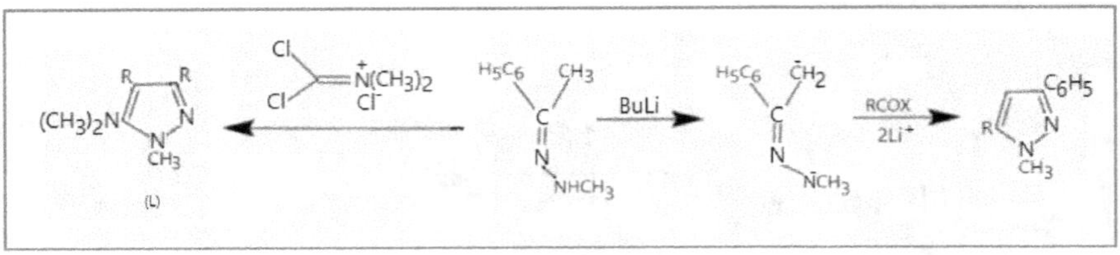

160

When cis-diaminomaleonitrile reacts with sulphur dichloride, it gives 3,4-dicyano-1,2,5-thiadiazole.

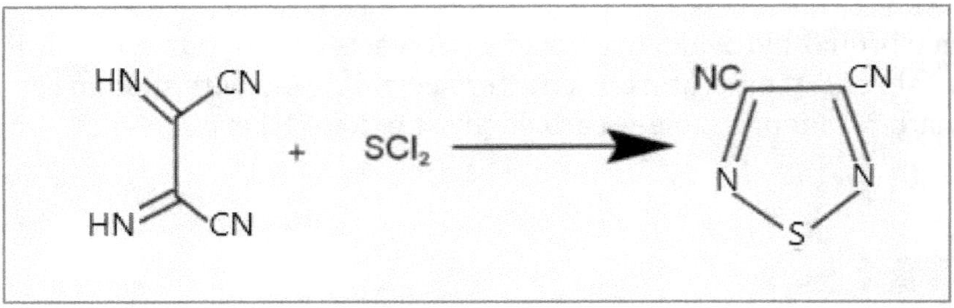

Anilines, adequately substituted at o-positions, reacts with phosgene-iminium chloride and yield benzoxazole, benzthiazole and benzimidazole. Imiazole can also be formed from 1,1-diphenylhydrazine (M).

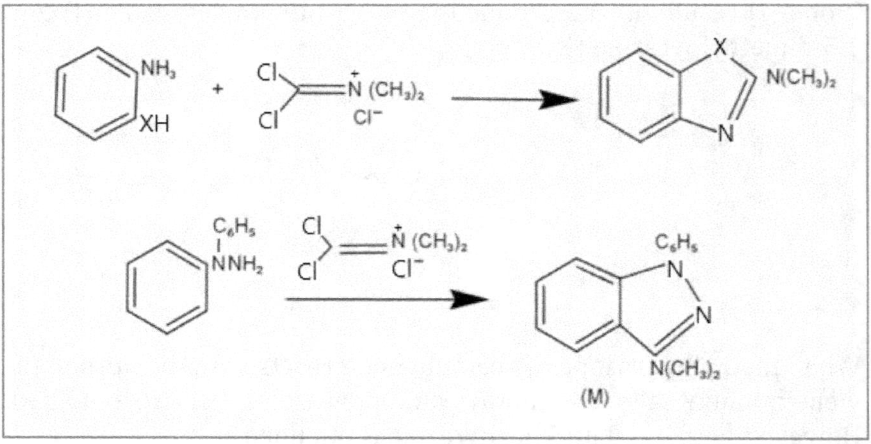

Likewise, phosgene iminium chloride also reacts with free base to give thiazolo[3,2-b][1,2,4]triazole derivative. Corresponding hydrazine give thiazole[2,3-c][1,2,4]triazole (N).

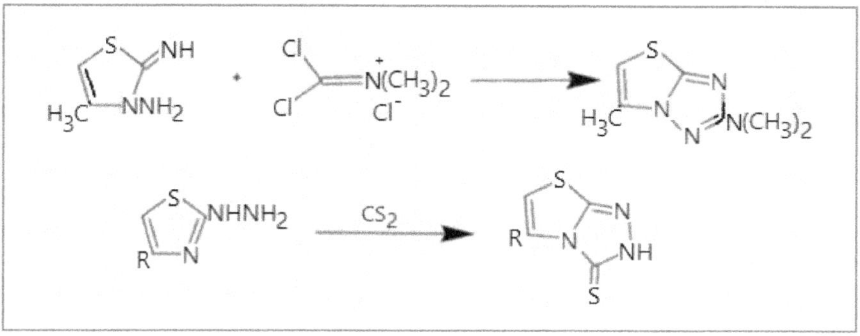

Benzene sulphonyl ester of mandelonitrile gives 4-aminothiazole.

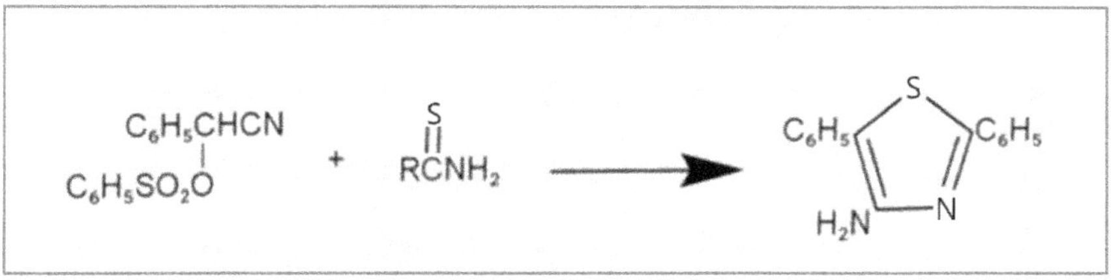

Variety of systems having nitrogen atom at the **fusion point** can easily be synthesized. 2-Aminobenzothiazole (O) reacts with phenylimino – oxalic acid dichloride (P) to yield 3-(phenylimino)-2-oxoimidazo[2,1-6]benzothiazole (Q) and with 2-mercaptobenzimidazole it gives 2-phenylimino-3-oxothiazolo-[3,2-a]benzimidazole (R).

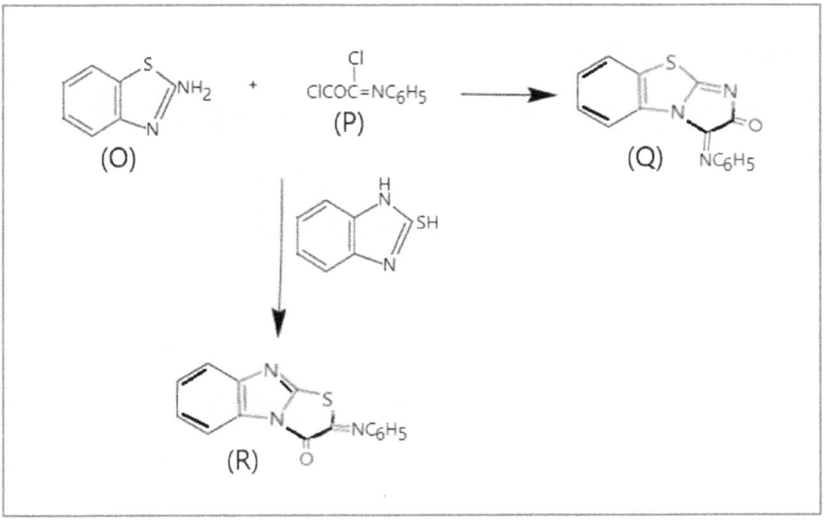

When trichloromethane sulphenyl chloride (S) reacts with 2-amino-pyridine in the presence of chloroform then sulphenimide intermediate results in ring closure product.

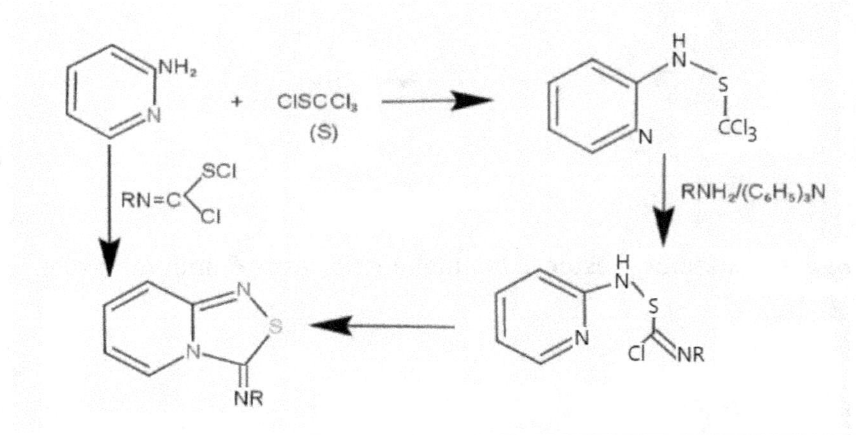

Dimethyl N-ethoxycarbonylthiocarbonimidate (T) reacts with hydrazine and hydroxylamine to give 1,2,4-triazolinone (U) and 1,2,4-oxadiazolinone (V) respectively.

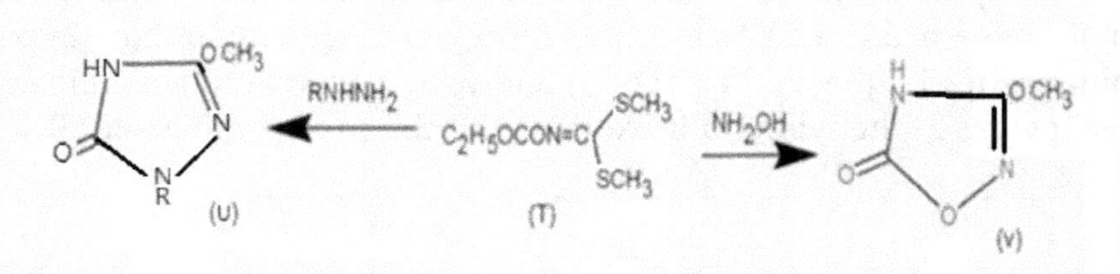

[1,3,4]-Selenadiazolo[3,2-a] pyridinium salt is obtained by the reaction of 3-chloro acrylaldehyde and 2-amino-1,3,4-selenadiazole.

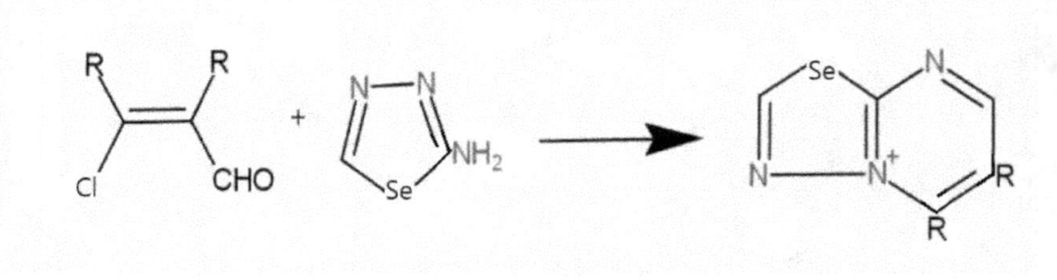

Benzylidene hydrazidines in mercuric oxide yields 4-arylamino-1,2,4-triazole.

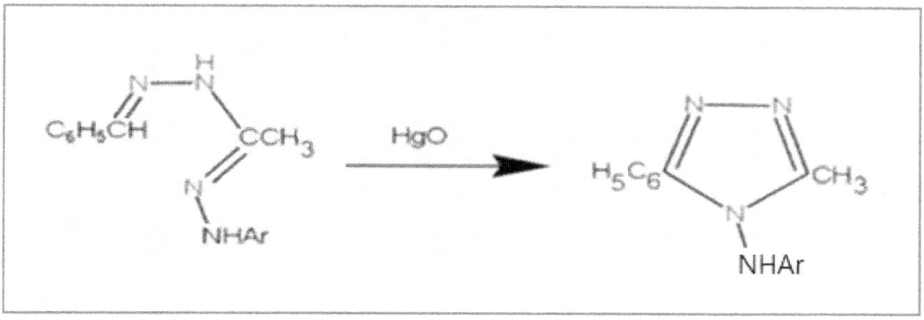

Substituted thiosemicarbazone react in two ways: with $Al_2O_3$ in chloroform it gives 1,2,4-triazoline-3-thione and with manganese dioxide in benzene thiadiazoline is obtained.

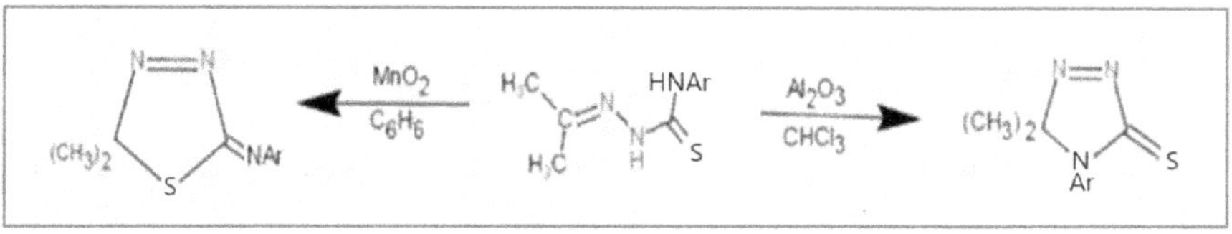

**Ring closure** of bis-hydrazone in the presence of mercurous acetate gives 1-amino-1,2,3-triazole derivatives.

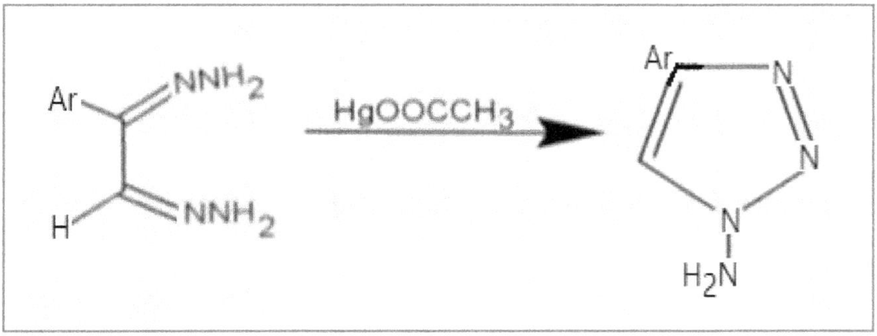

**Oxidation** in the presence of hydrogen peroxide, synthesizes 5-amino-4-cyano-3-methyle isothiazole.

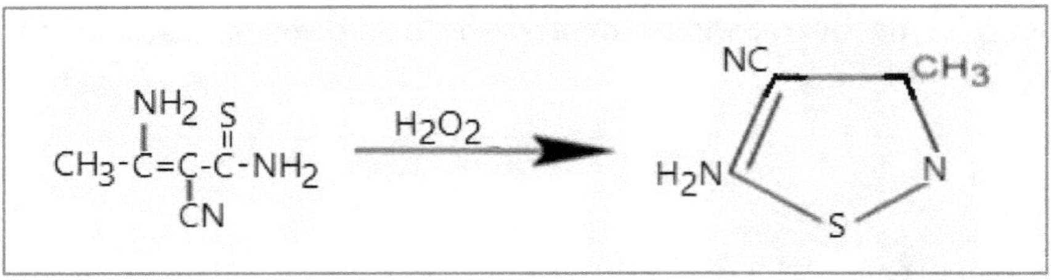

164

Ring closure of 2-acetamido-5-thiazolylthioglycolic acid in the presence of phosphorus oxychloride gives 2-acetamidothieno[3,2-d] thiazolin-4-one.

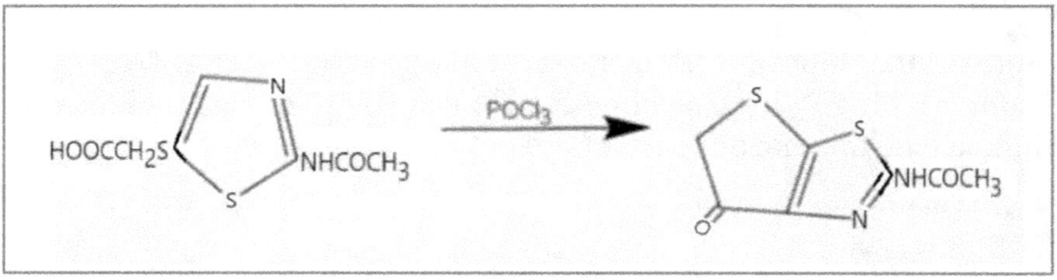

Ring closure via nucleophilic addition gives 4-amino-5-cyanoimidazole.

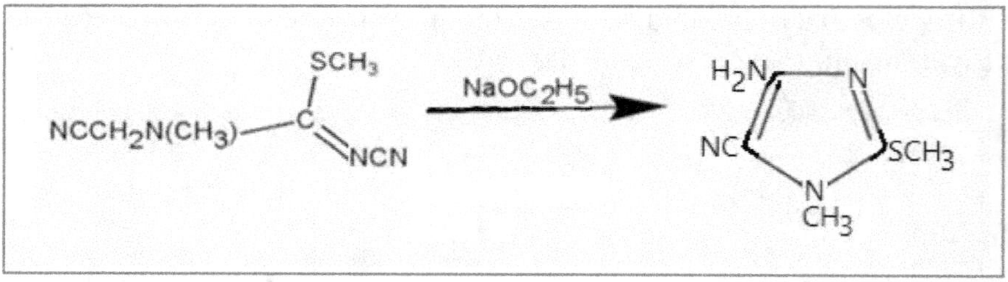

Nitrilimines (W) yield substituted pyrrazoles and triazoles.

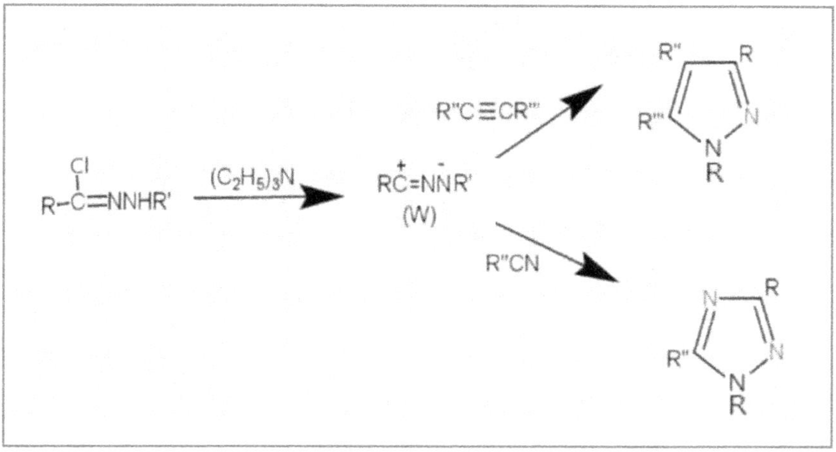

**Decarboxylation** with subsequent ring closure results in the formation of heterocyclic ring when 12-benisoxazole is treated with phenylhydrazine.

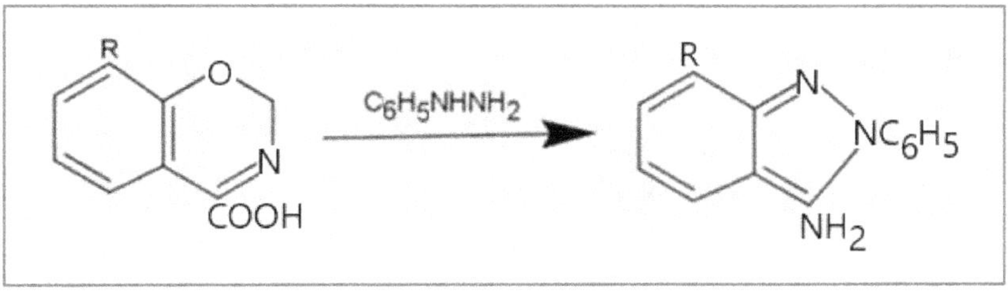

1,2,4-Oxadiazoles are obtained when 1,3,5-oxadiaziniumsalt (Y) reacts with hydroxylamine while with hydrazine it forms 1,2,4-triazole derivatives. In the same way, 1,3-oxazinium salts (X) and hydrazine give pyrazoles.

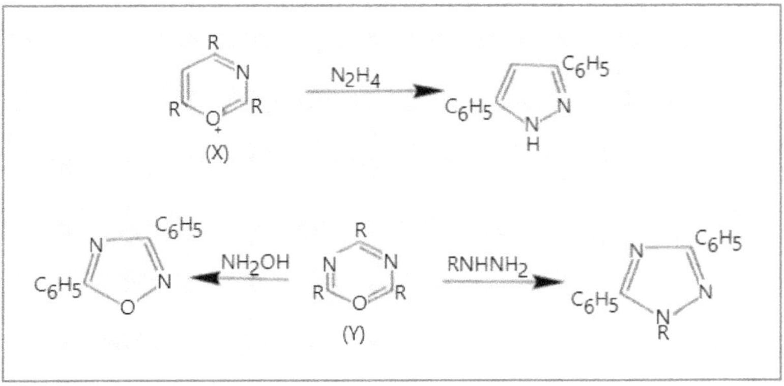

Pyrazolo[3,4-d] pyrimidines are produced by **oxidative cyclization** of methylated 6-(benzylidenehydrazino)uracil.

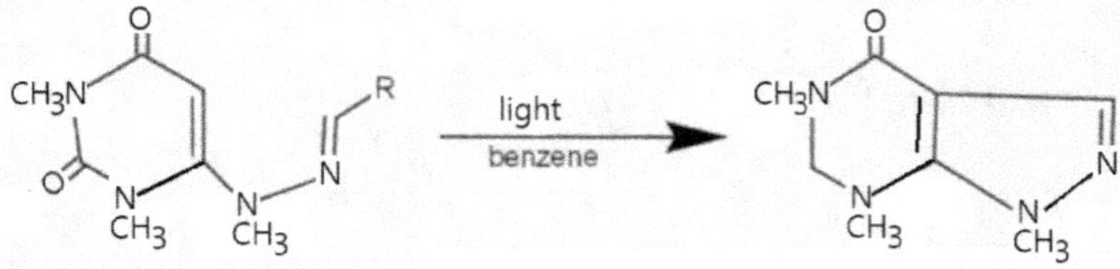

Conversion of diphenylketene of sulphimide results in imidazo[1,2-a]pyridine-3-one(Z).

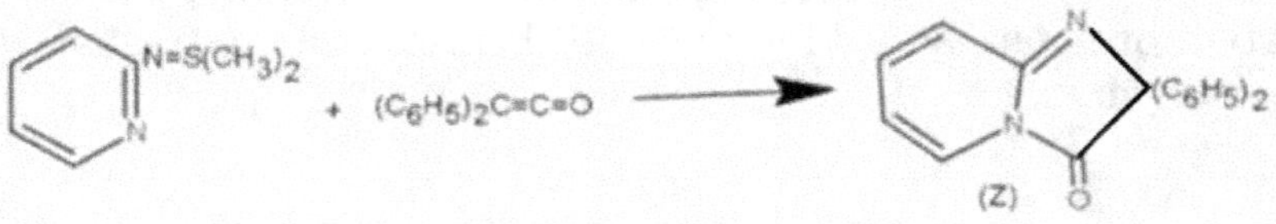

## 7.3 GENERAL CHEMICAL PROPERTIES

**Photochemical fragmentation** of 1,2,3-thiadiazole (A) gives thiirene (B) that can further give products with alkynes.

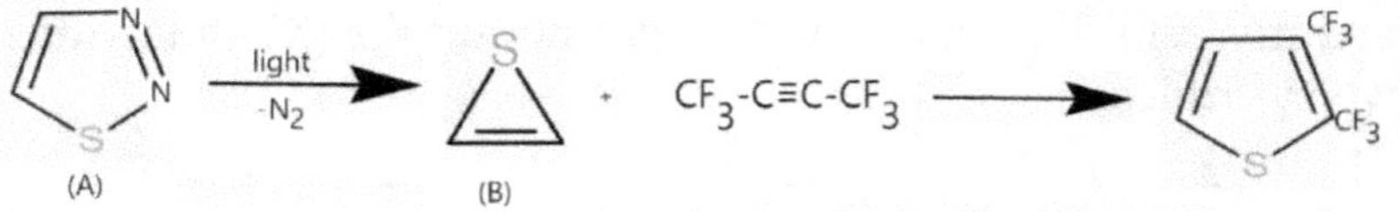

Multiple heterocyclic compounds yield 3-phenylthiarizine (C) as intermediate which results in the formation of simpler products.

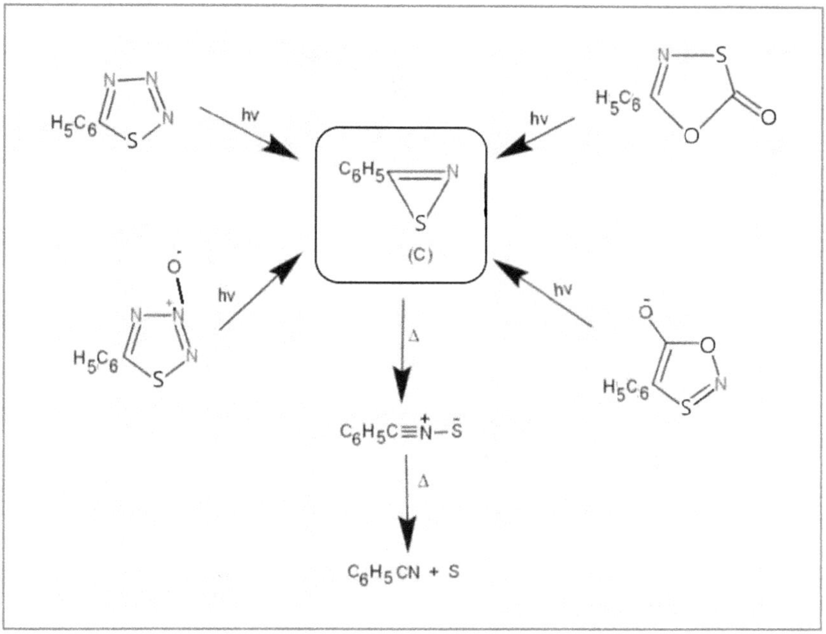

Tetrazoles on photolysis yields diaziridine.

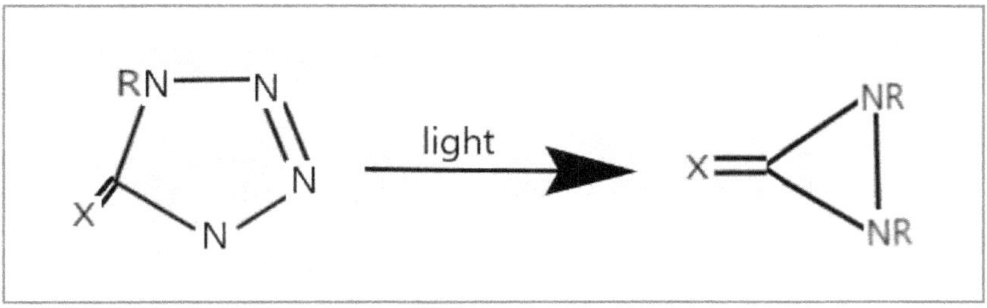

**Photochemical conversion** of ppyrozolin-5-one to imidazolones and open-chain products is easy to obtain.

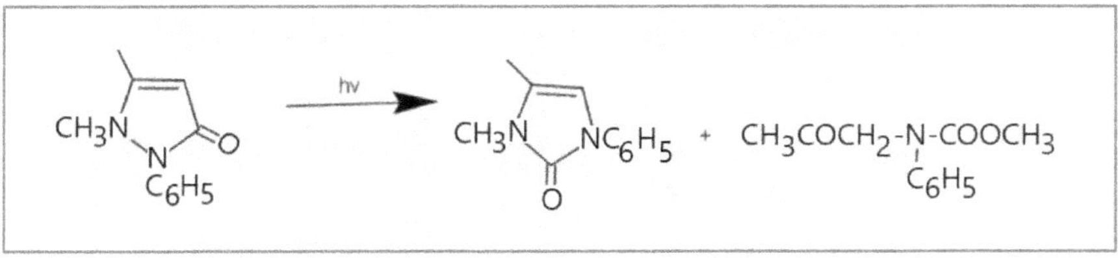

**Regioselective alkylation** of imidazole gives sterically less favoured derivatives.

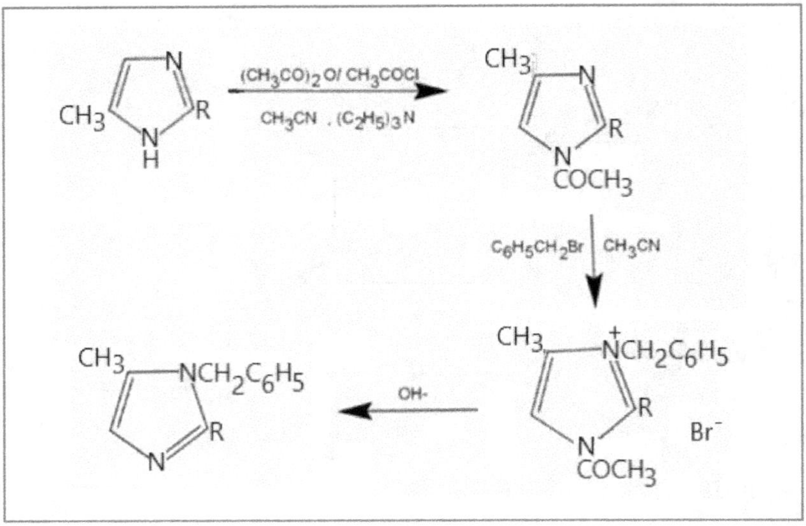

168

**Oxidation** of imidazole rings by hydroperoxide in presence of light gives diarylbenzaimidines.

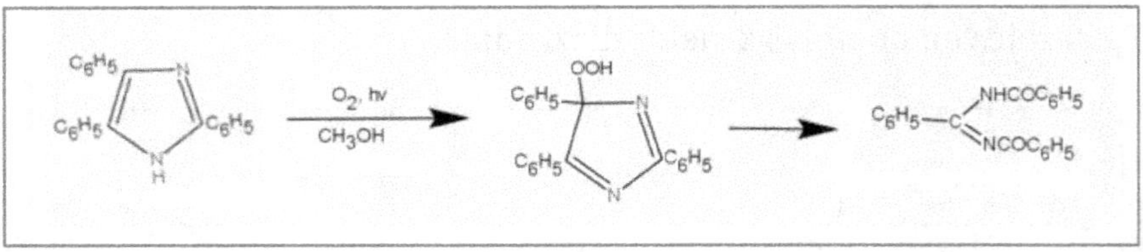

**Ring opening products** as benzil α-monoxime benzoate (b), are obtained with ozone.

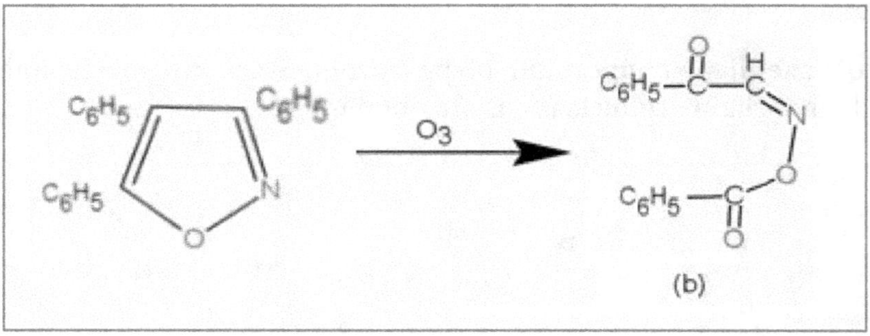

Organolithium compounds bring about substitution at position-5 in 1,2,4- oxadiazole(E).

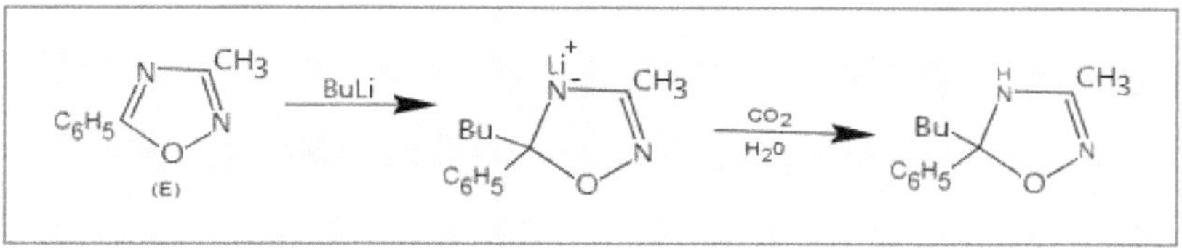

**Reduction** of oxygen containing heterocyclic compounds is brought about readily.

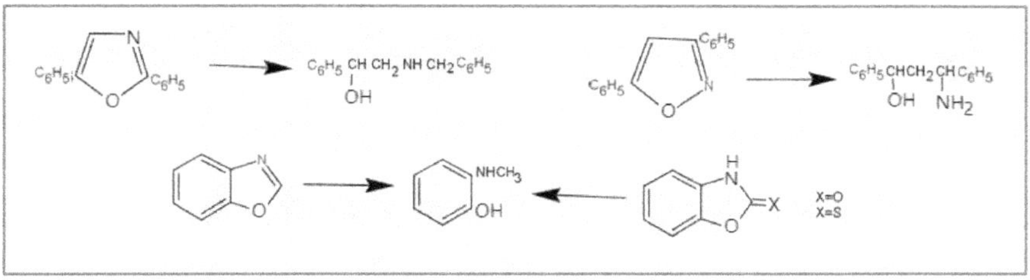

**N-** imines are obtained when 1-alkyl-1,2,4–triazoles react with nitrenes.

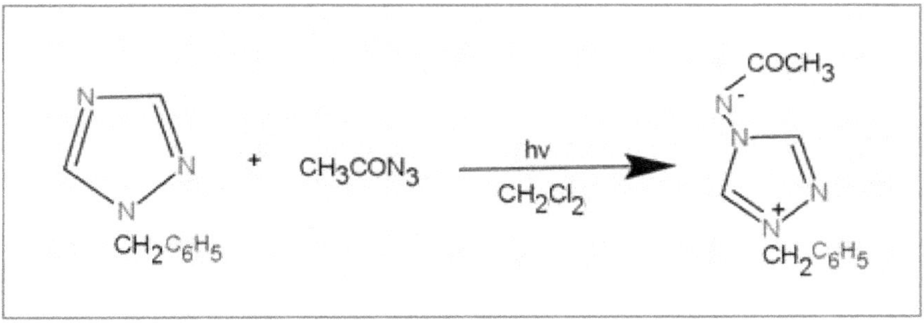

Isoxazoles get easily reduced to give ring fission products.

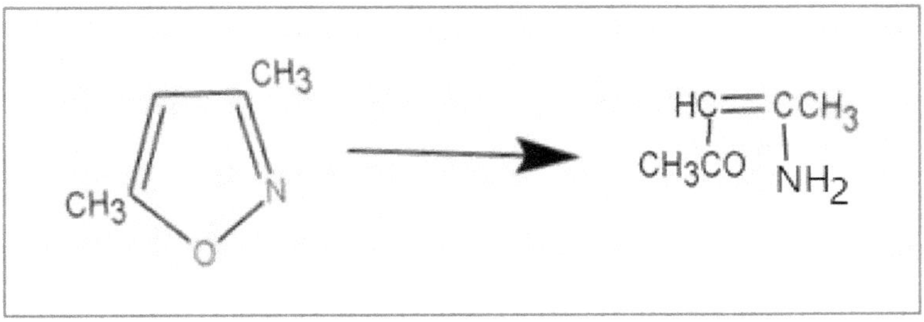

Various reduction products are obtained from 1,2–benzisoxazoles.

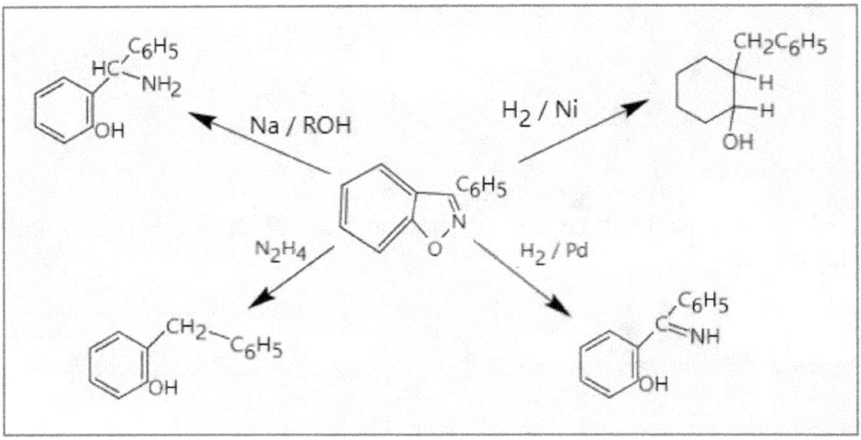

**Cleavage** of heterocyclic rings in oxazoles takes place in the presence of catalyst.

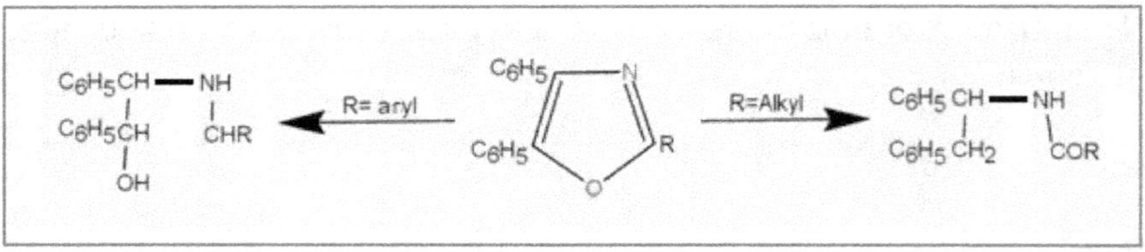

1,2,5- Thiadiazoles show **reductive cleavage** by zinc in acid or sodium in alcohol.

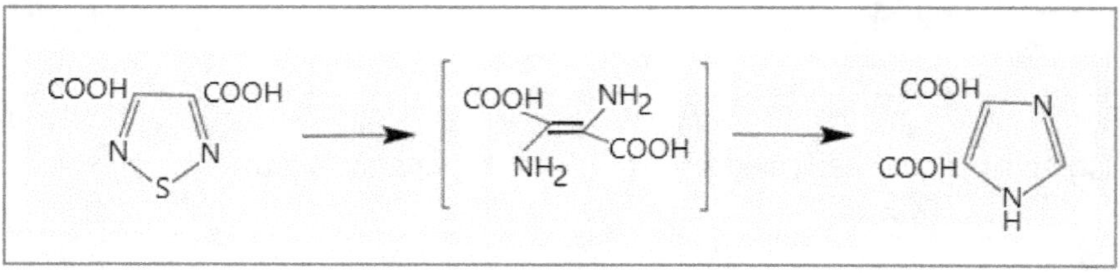

Azolecarboxylic acids and its derivatives show a variety of reaction with various reagents.

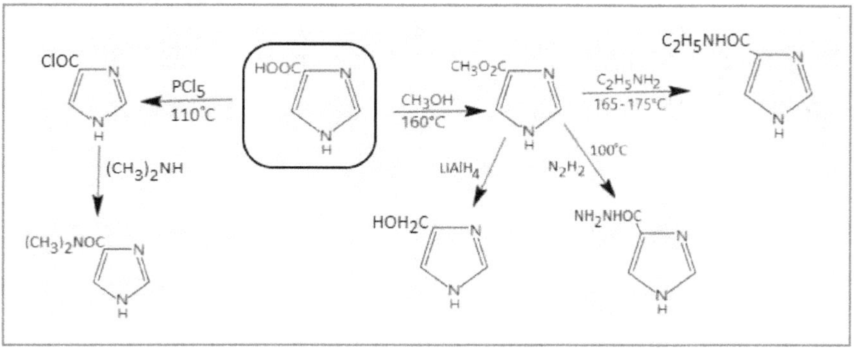

## 7.4 PYRAZOLE AND ITS DERIVATIVES

Pyrazoles belong to azole family i.e., five membered rings containing only carbon and nitrogen and they range from pyrazole to pentazole.

## 7.4.1 SYNTHESIS

Azines of hexafluoroacetone through a series of reactions results in substituted pyrazoles.

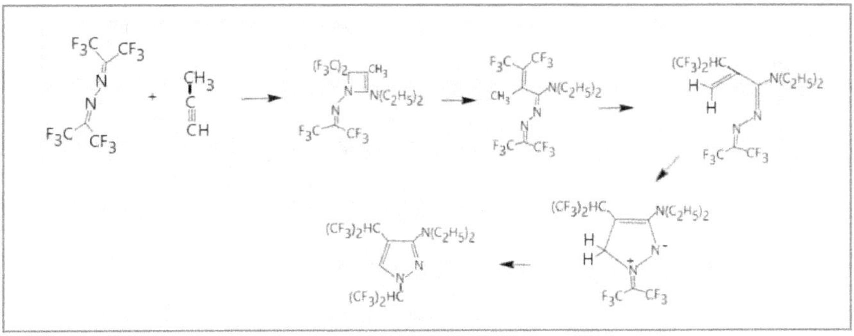

2,2 – Diindazolyl derivatives are obtained from hydrazines.

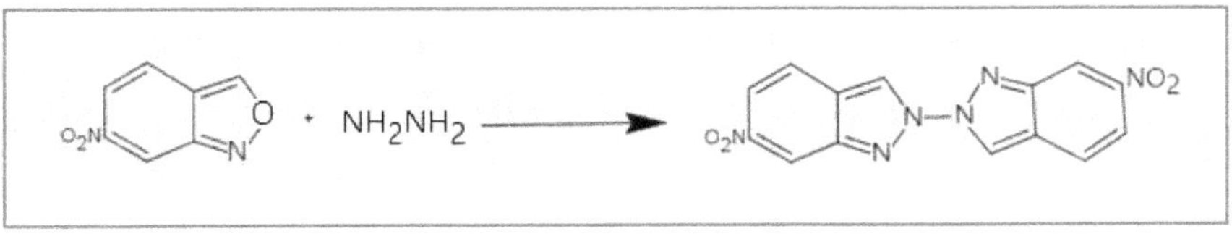

Thionyl chloride reacts with δ-keto acids to give pyranopyrazole which opens up with hydrazine. Later on, it forms pyrazolodiazepine or pyrazolopyridine.

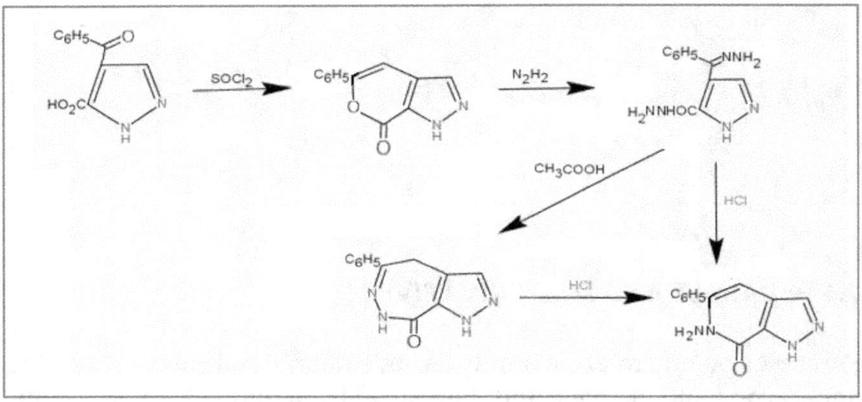

Substituted anthranilic acid, on getting treated with phosphorus oxy diboride results in the formation of a new pyridine ring. However, it's another isomer forms pyrazolo quinazoline (B). The tri aryl pyrazole on photolysis cyclizes with loss of hydrogen chloride.

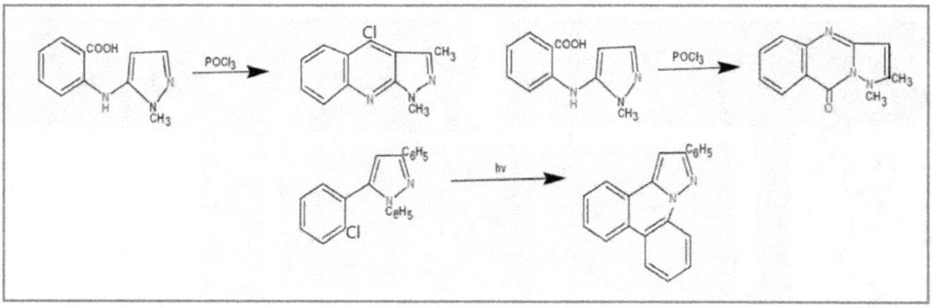

3-Aminopyrazole reacts with various esters and nitriles to produce many compounds.

## 7.4.2 PROPERTIES

**Rearrangement** or rings opening products are formed depending on the substituent at position – 4; however, N-N bond is disintegrated in all the cases in the presence of light.

174

Pyrazole is brominated with bromine in chloroform and bromine in acetic acid.

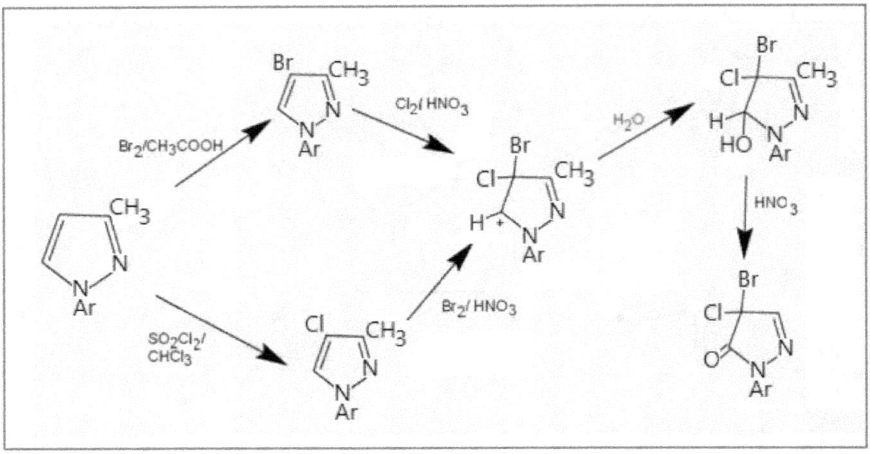

 3,4,5-Trimethyl pyrazole reacts with carbene in basic as well as neutral medium. Under basic condition (CHCl$_3$-C$_2$H$_5$ONa), 4-dichhloro-methyl–3,4,5-trimethyl iso-pyrazole is obtained which finally shows **ring expansion** into ethoxy-methyl pyridazine. However, in neutral medium, pyridazine and pyrimidine is obtained.

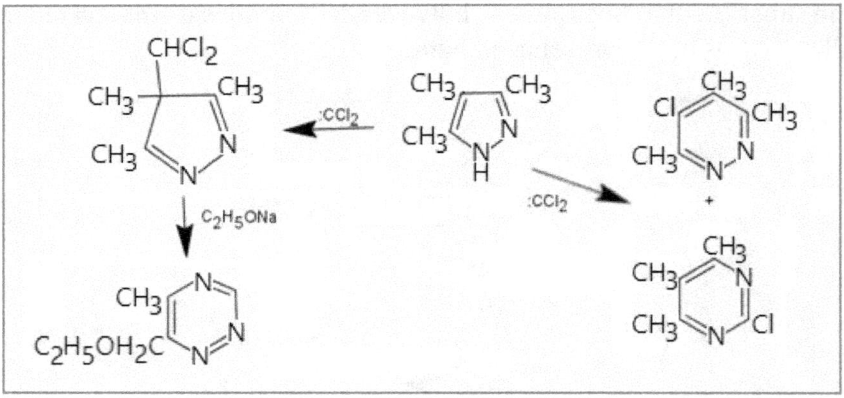

3-Methyl-1-phenyl-2-pyrazolin-5-one show many reactions in various conditions.

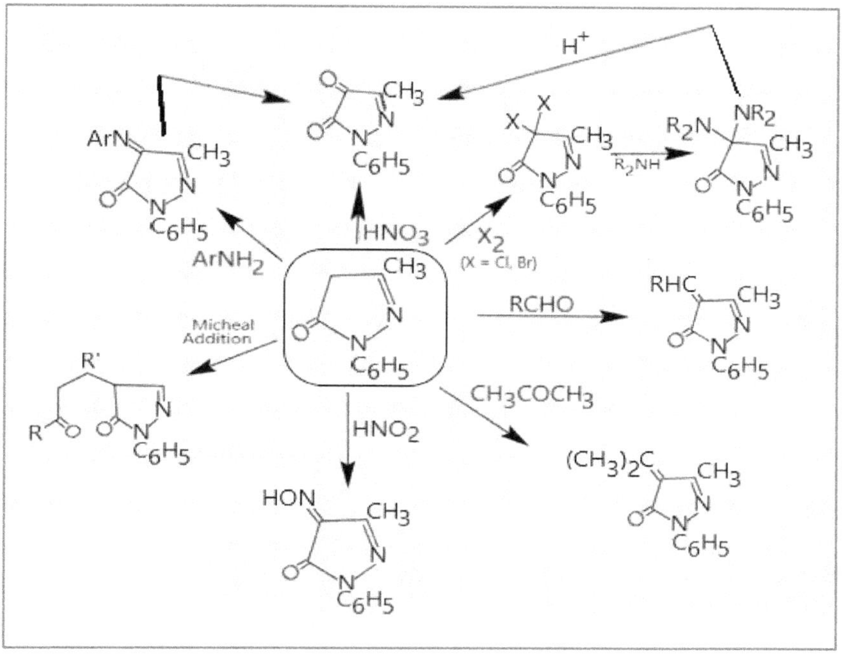

Dimethyl sulphate and sodium hydroxide show **N - methylation** in pyrozolo- pyrimidinone while diazomethane gives mixture.

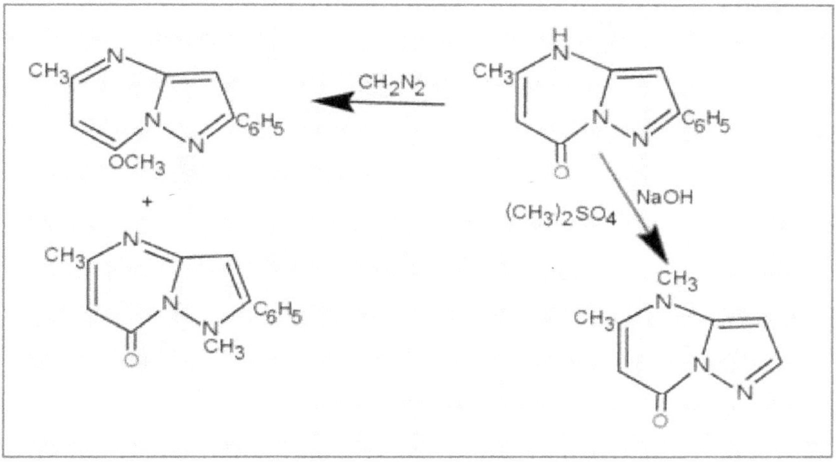

**Nitration** of pyrazolo[1,5-a] pyrimidine gives 3-nitro and 6-nitro derivatives in different conditions.

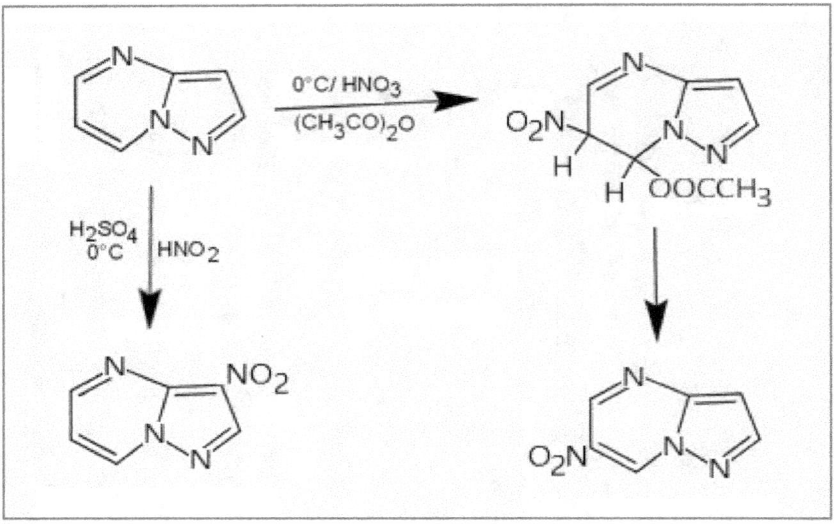

176

**Nucleophilic displacement** of pyrazole[3,4-d]pyrimidine occurs with difficulty.

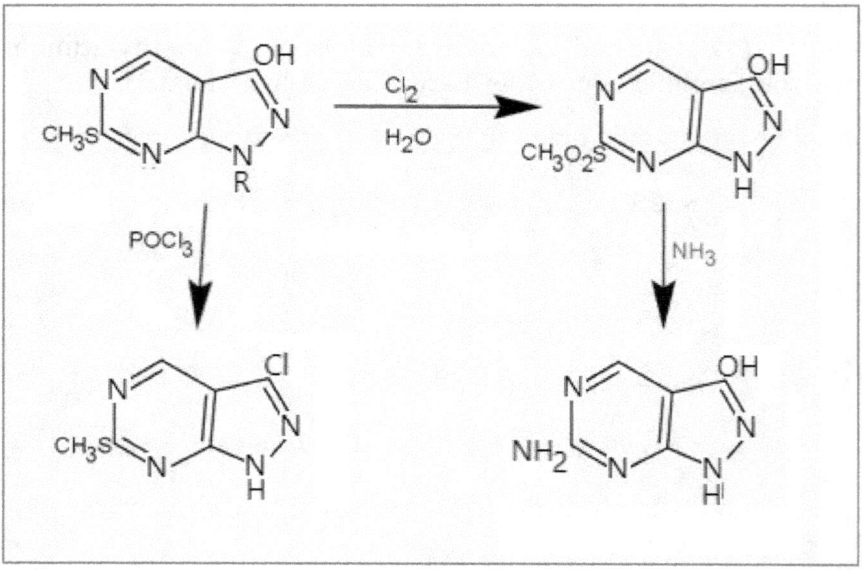

**Reduction** in the presence of palladium or platinum first removes substituted halogen and then saturates the ring.

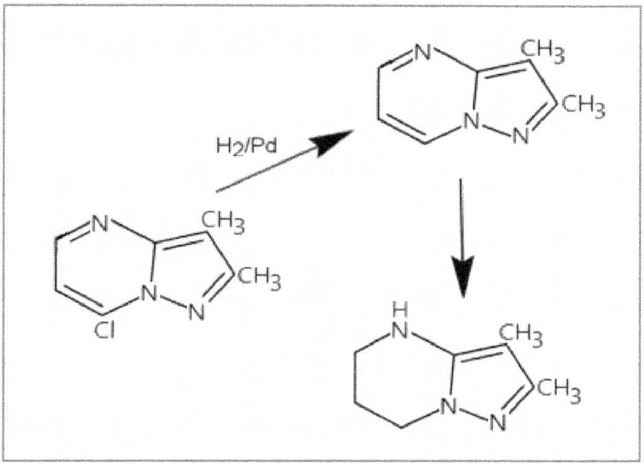

## 7.4.3 APPLICATIONS

(i) 1,4–Dihydro[1]benzothiopyrano[4,3-c)-pyrazole shows antimicrobial activity.

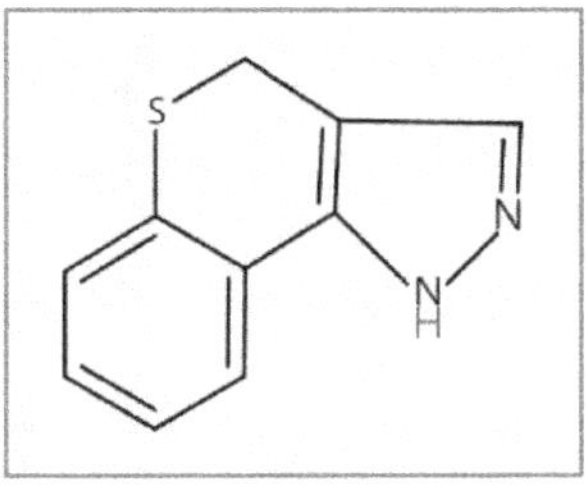

(ii)    **Forbisen**, 2,2', 3,3'-tetramethyl–1,1'-diphenyl-4,4'-bi-3,3'–pyrazoline-5,5'-dione, a by-product in the manufacture of antipyrine is used in bovine anaplasmosis.

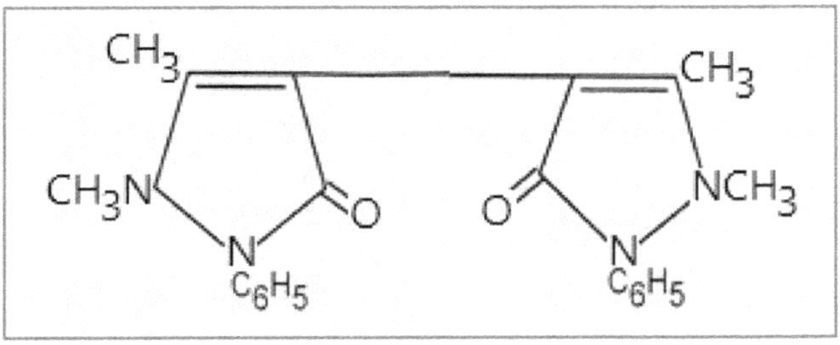

178

(iii)    **Muzolimine**, 1-substituted 2-pyrazolin-5-one derivatives is used as an active diuretic.

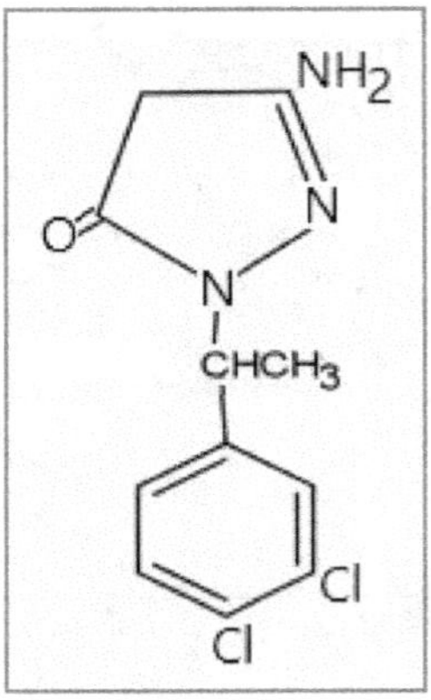

(iv)    **Dimetilan**, dimethylcarbamic acid 1-[(dimethyl amino)carbonyl]-5- methyl -1H- pyrazol-3-yl ester has been found to have insecticidal properties.

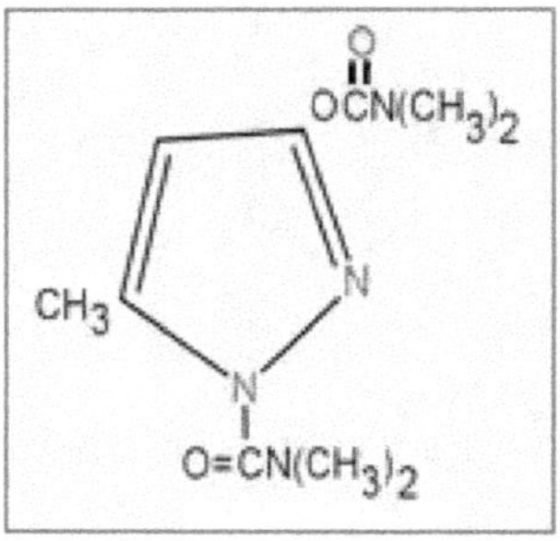

(v)    Dyes of 5-pyrazolone derivatives are very widely used, such as: **Rosathrene orange (a), Pyrazol orange (b), Polar yellow (c), Eriochrome Red B (d), Vulcan fast red GF(e)** etc.

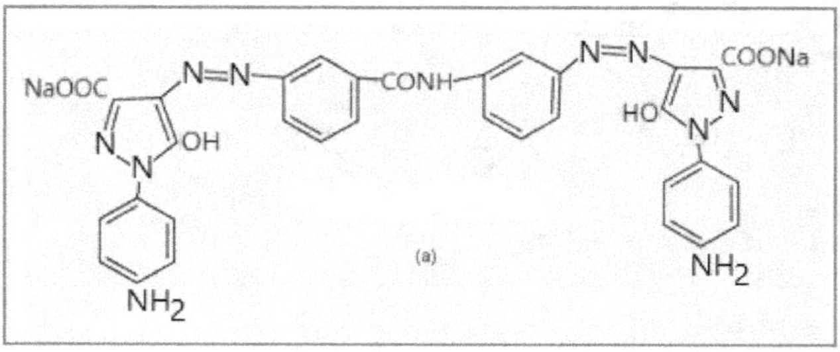

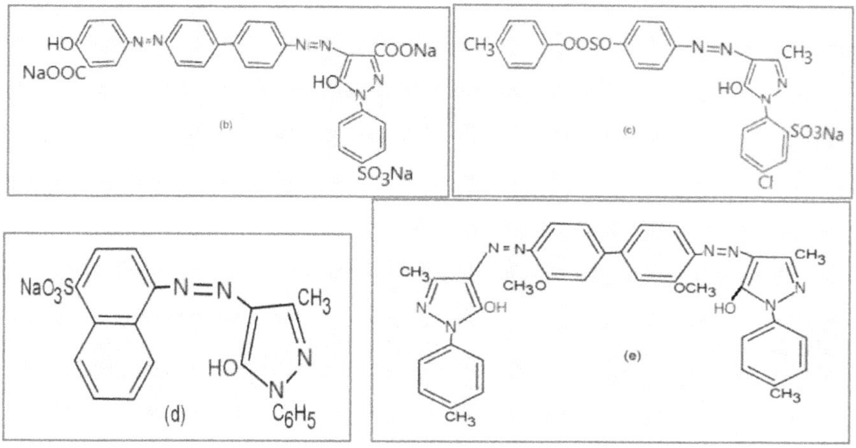

## 7.5 IMIDAZOLE AND ITS DERIVATIVES

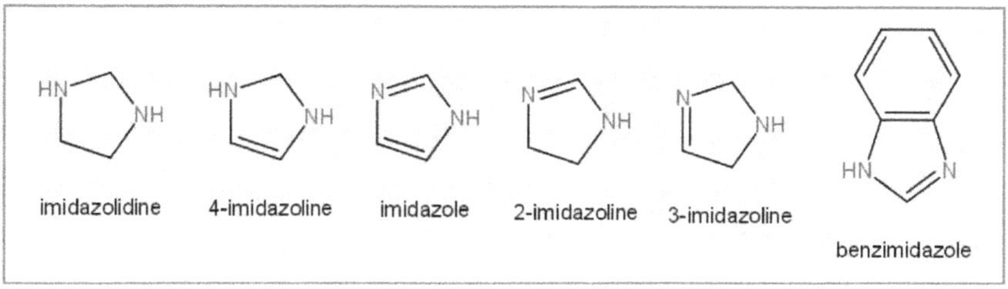

## 7.5.1 SYNTHESIS

In reductive environment, N, N–diethylaniline is converted to benzimidazolinone (A) in presence of transition metal complexes. However, with N, N- diethylanilines, benzimidazole (B) is obtained.

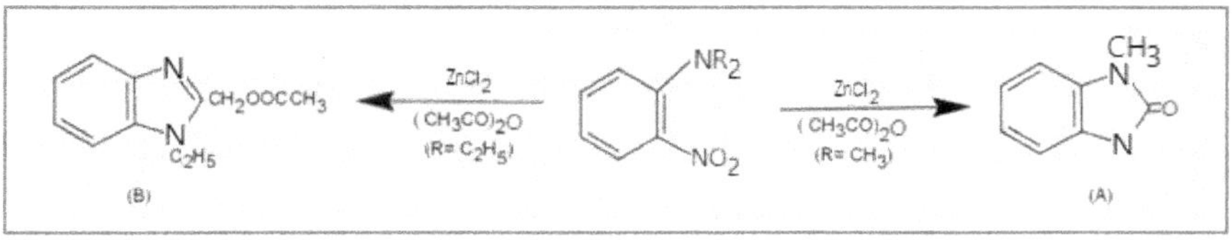

Imidazopyridine is obtained by a method of synthesis of bicyclic product.

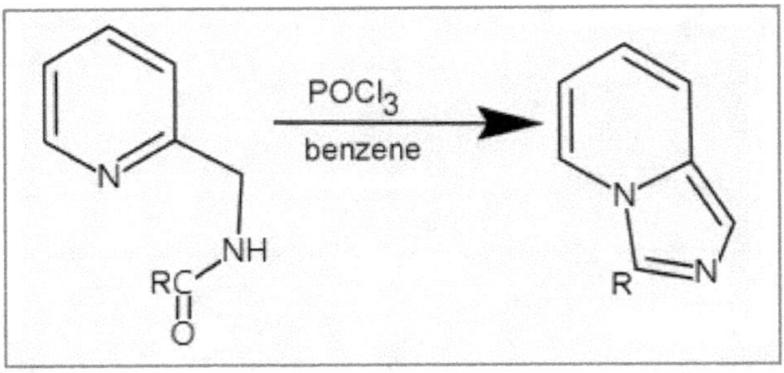

180

**Ring closure** of cyanide carbon to imidic nitrogen forms a series of mesoionic imidazole (E) through 4-amino imidazolium salt (C and D).

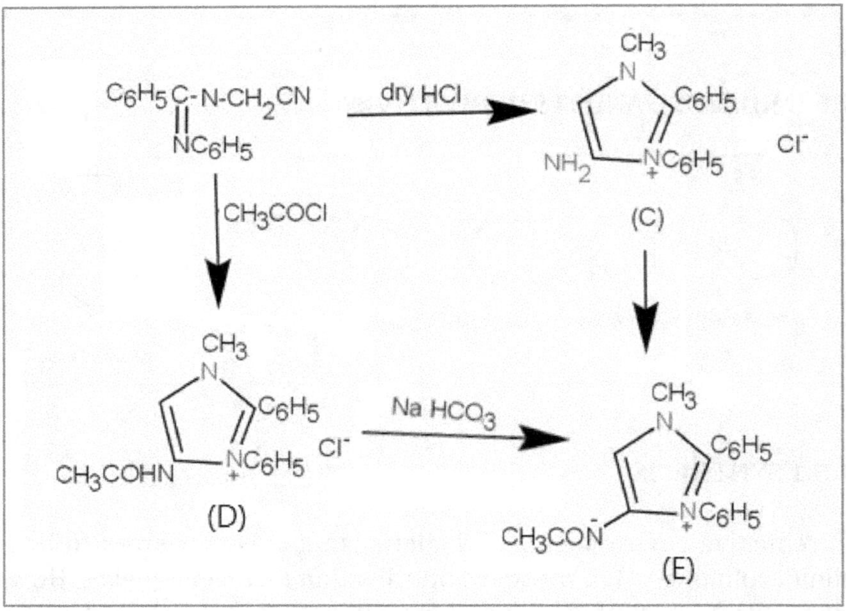

2-Substituted benzimidazoles can be obtained by a variety of methods from O–arylenediamine.

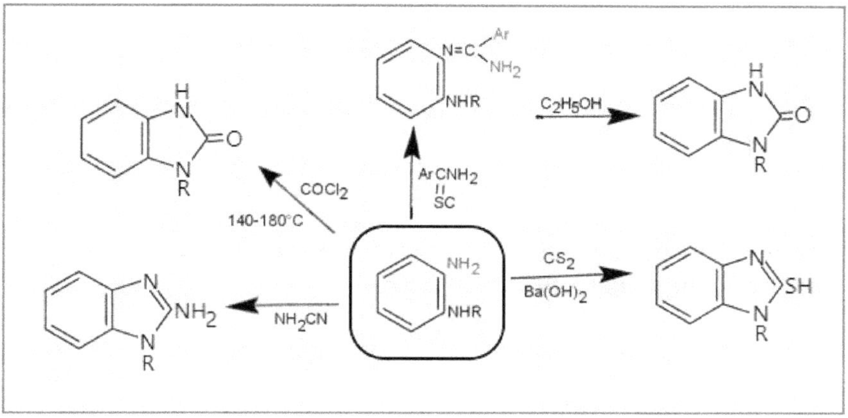

Pyruvaldehyde (F) can be used as starting material for the formation of 4-methyl and 2,4-dimethylimidazole.

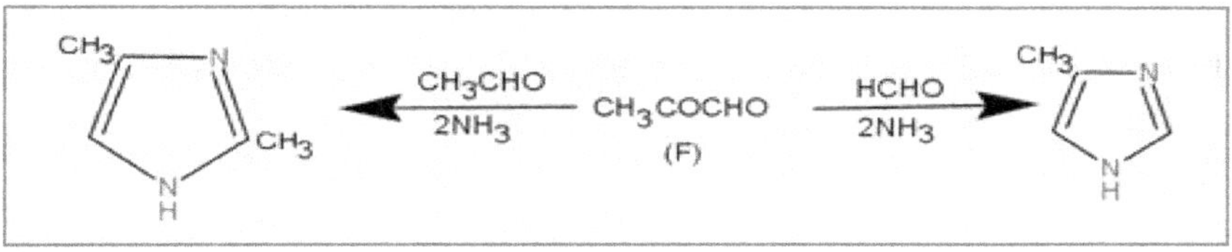

## 7.5.2 PROPERTIES

2,2 – Dimethyl-2H-benzimidazole-1,3-dioxide undergoes fragmentation on UV irradiation.

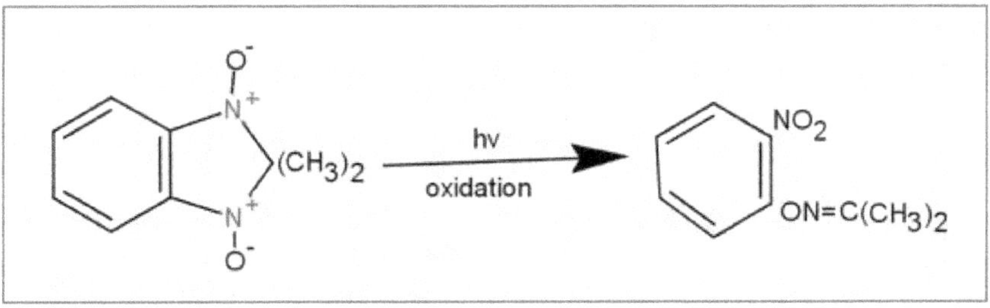

**Thermal rearrangement** occurs in 1,4-dinitroimidazole at 140°C in chlorobenzene to form various nitro-products.

**Photocyclization** of trans-1- styryimidazole takes place when it gets transformed into cis-isomer at 300nm.

**Thermolysis** of 2,4,4-triaryl-5-methylthio-4H–imidazoles produces 1,1'-biimidazolyl.

Hydroxide ion induces ketonic group at position-2 in benzimidazoles.

2-Pyridinium compounds (B) are obtained when cycloheptimidazolin-2(1H)–one reacts with phosphoryl chloride in pyridine.

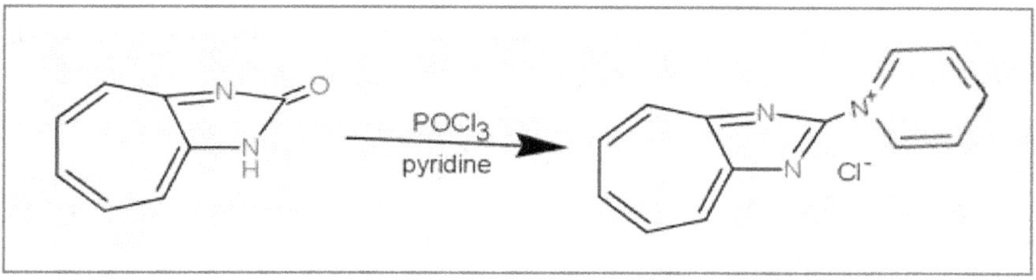

Phosphoryl Chloride also changes 4,5-diphenylimidazolin–2-one into 2-chloro product.

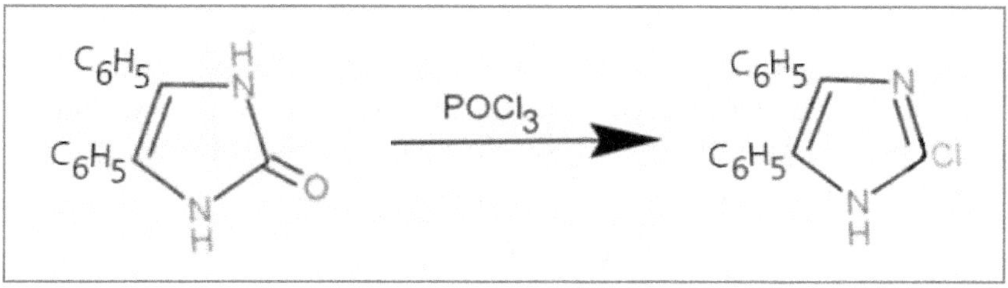

**Halogeno-denitration** takes place in imidazole rings substituted by more than one nitro groups with position–2 displacing first and position–4 displacing last.

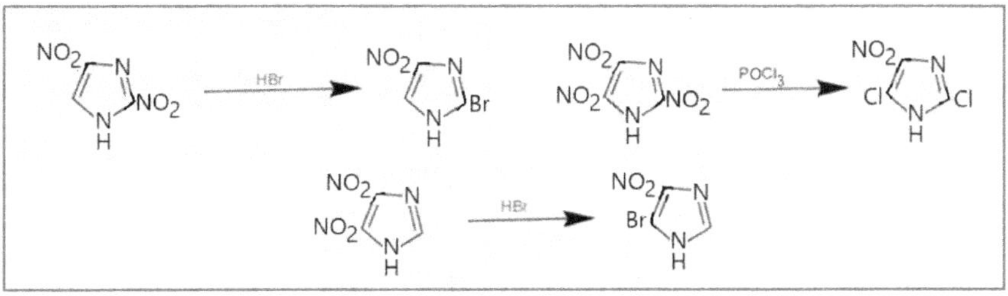

With benzophenone, 1,2–dimethylimidazole adds up diarylketone at position–2 methyl group. However, 1- benzylimidazole adds up at C-1 methylene.

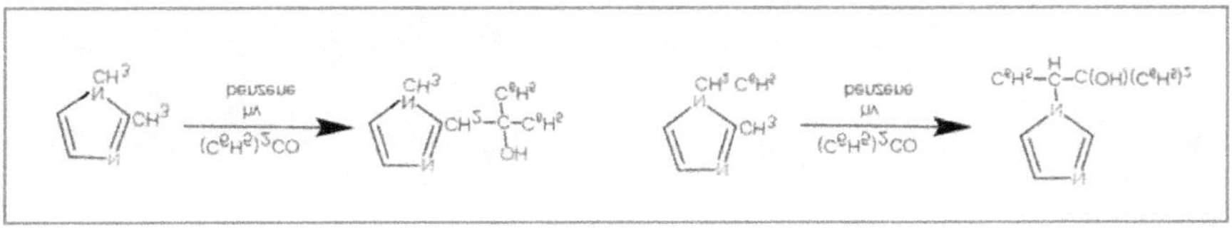

184

5-H- imidazolin-4- one (C) undergoes **Schotten-Baumann acylation** to give product on ketonic oxygen. N-acylation takes place on changing the reaction conditions.

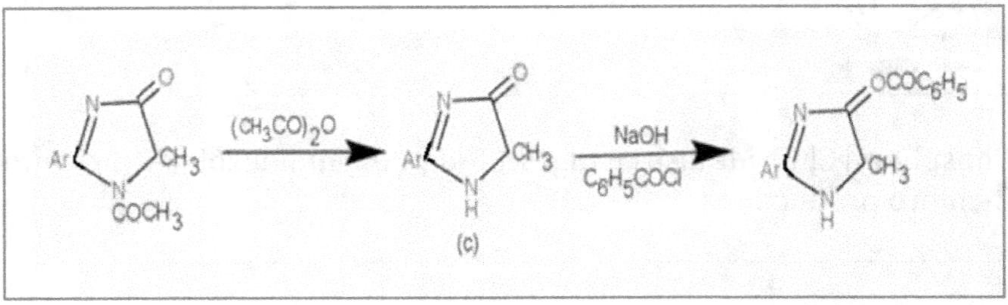

When imidazole–4,5–dicarboxylic acid is heated with acetic anhydride, then compound (D) is formed which yields a monoester (E) with ethanol and hydroxide (F) with hydrazine.

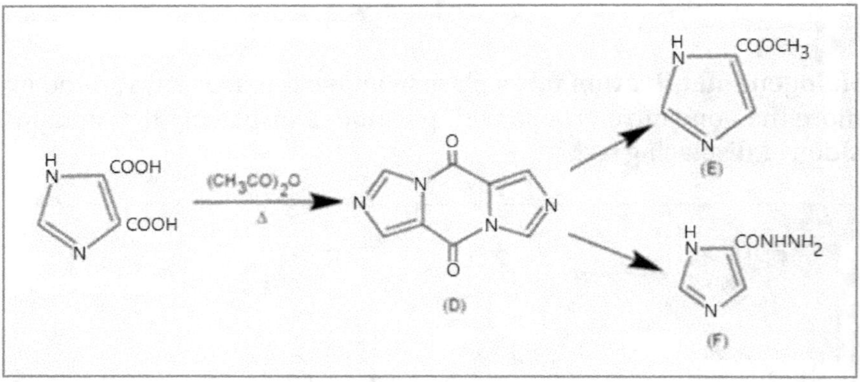

A variety of condensation reactions are shown by 2–ethynyl–1–methyl benzimidazole.

2 – Azidobenzimidazole shows a variety of reactions.

**Photorearrangement** of 2–aryloxybenzimidazole forms 2–(2– and 4– hydroxyphenyl)benzimidazole.

S – Alkylation is shown by **addition** of alkynie esters to benzimidazole – 2 – this one.

### 7.5.3 APPLICATIONS

(i)     Amongst natural imidazoles, amino acid **Histidine** (a) is especially important; while imidazo-pyrimidine, **Adenosine** (b) is also of great importance.

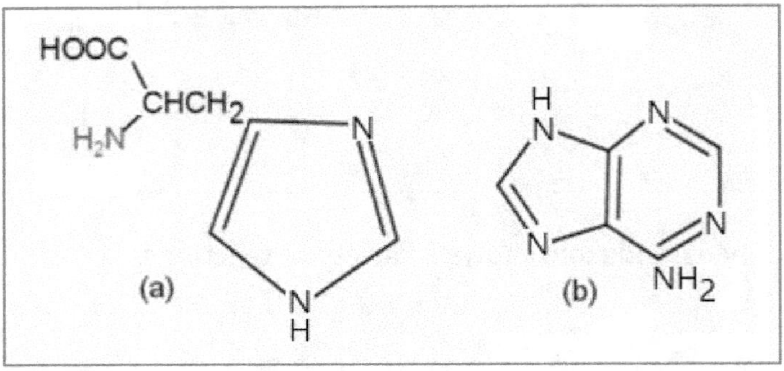

(ii)     **Biotin** (c) is an essential growth factor & **Piscol** (d) is a valuable vasodilating drug.

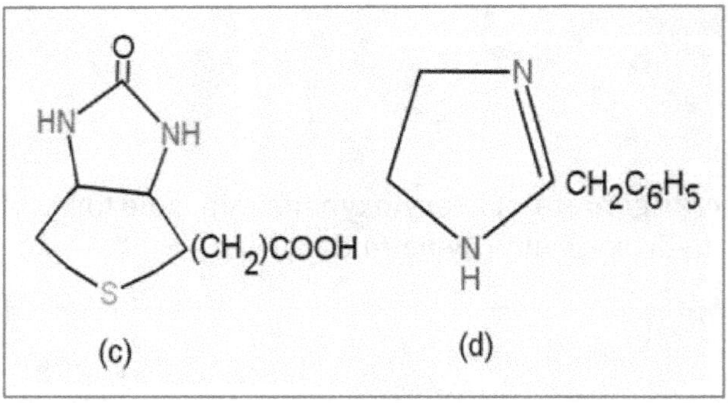

## 7.6 MISCELLANEOUS COMPOUNDS & THEIR DERIVATIVES

### 7.6.1 PURINES

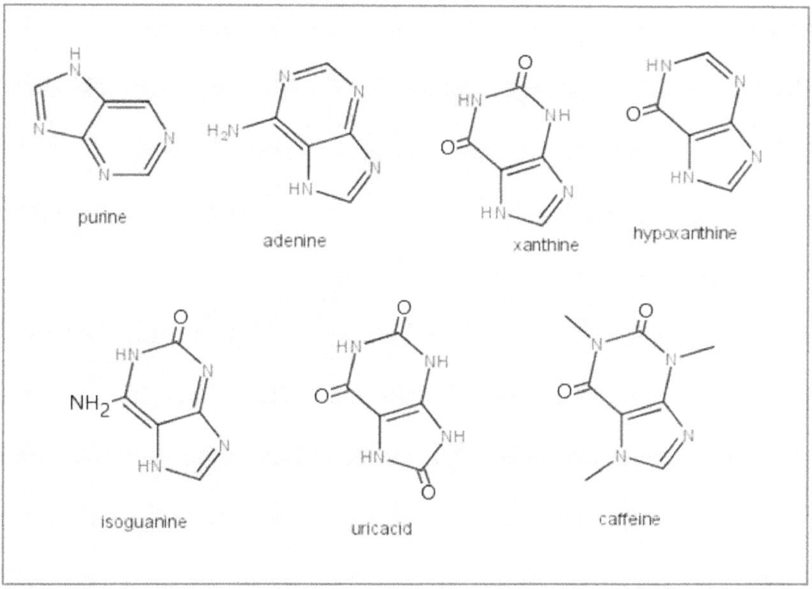

purine

adenine

xanthine

hypoxanthine

isoguanine

uricacid

caffeine

## SYNTHESIS

**Vilsmeier – Haack reagent** (Dimethyl formamide, DMF and phosphoryl chloride) is used to obtain 8–dimethylaminotheophylline.

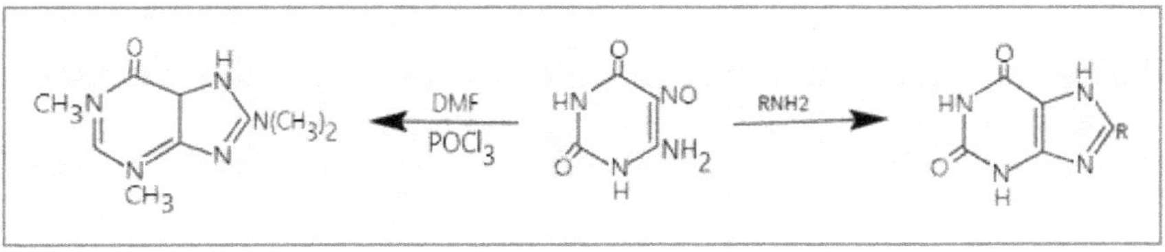

Aminothioamide, on being fused with urea yields 6-thioxanthine while fusion with amino imidazoleamidine gives isoguanine.

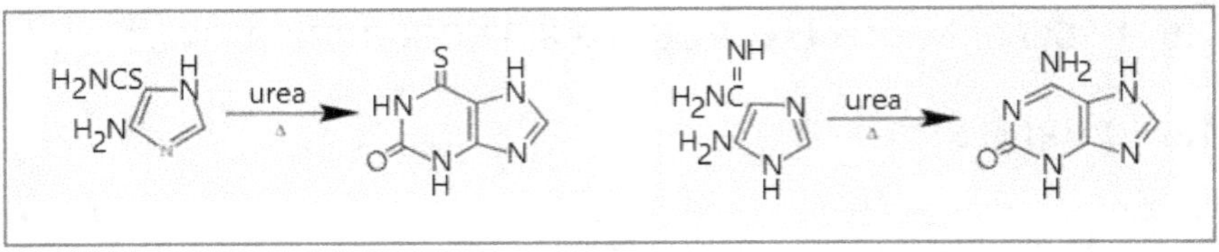

**Caffeine** is obtained when trimethyl imidazole reacts with ethylchloroformate in aqueous sodium bicarbonate.

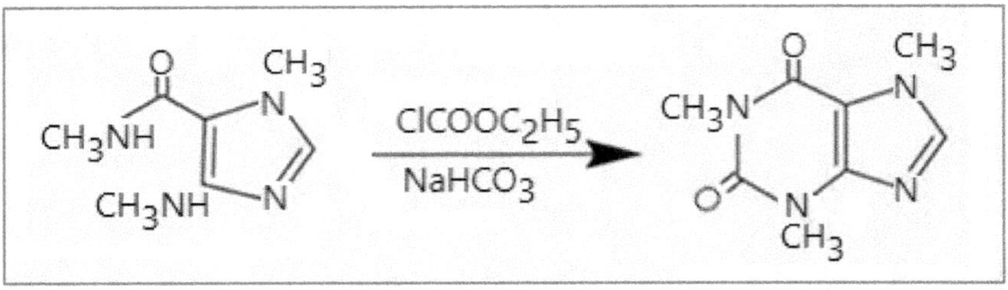

Derivative of 9–methyl adenine is produced by reaction between methylamine and oxazolo derivative.

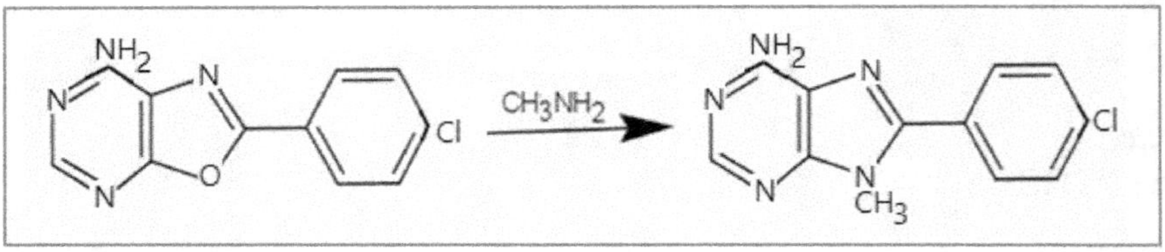

## PROPERTIES

**Methylation** take place in 1 – methylxanthine to produce caffeine in the presence of dimethyl-sulphate in alkaline medium.

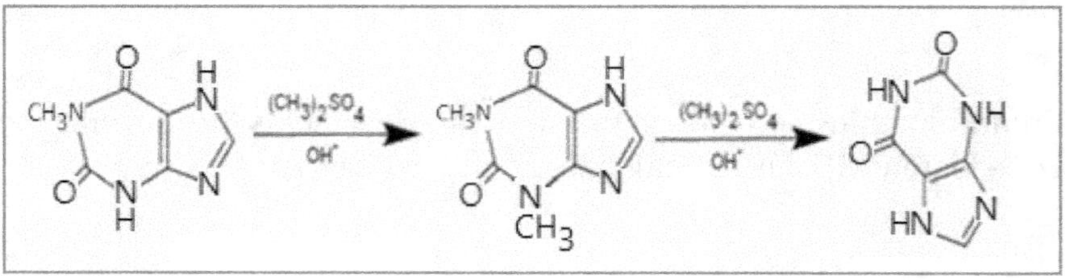

Substituted adenines are produced by **intermolecular free radical process** by heating.

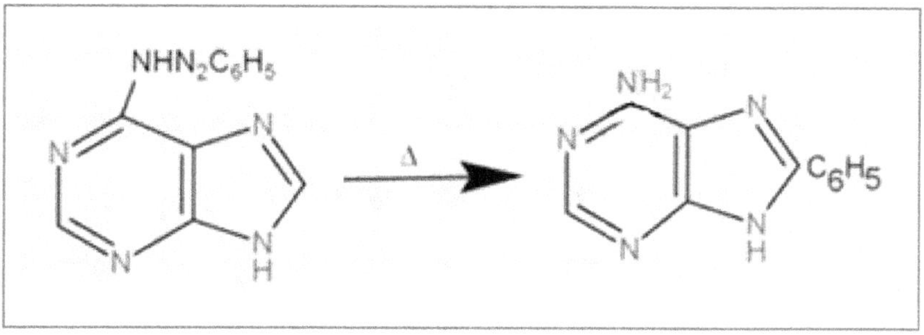

1 – Methoxy adenine shows migration of groups and results in new products.

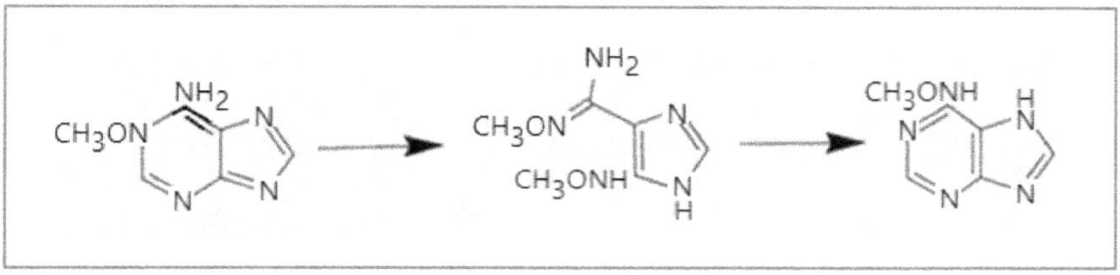

## 7.6.2 TRIAZOLES

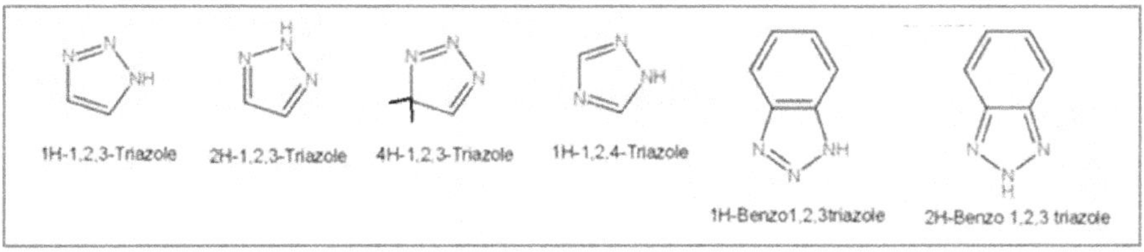

## SYNTHESIS

Linear triazines (A) cyclize to form 5–amino–1–aryl–1,2,3–triazoles via Lewis base catalysis.

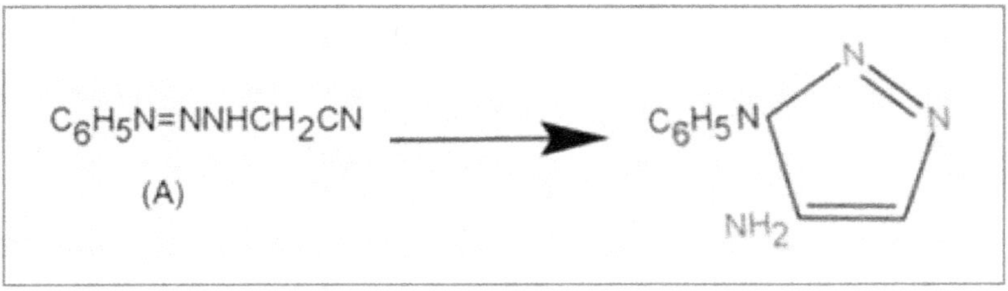

The triazoline is obtained when azides containing electron withdrawing substituents add to allenes.

**Diazotization** of 4–amino–3,5–dimethyl isoxazole (B) results in the formation of triazole.

**Cyclization** of 1,2-azoaminobezene results in the formation of 2–phenyl-2H-benzotriazole (C).

In the same way, o–nitrohydrazine derivative **cyclizes** to yield 1–hydroxy-6-nitro benzotriazole (D).

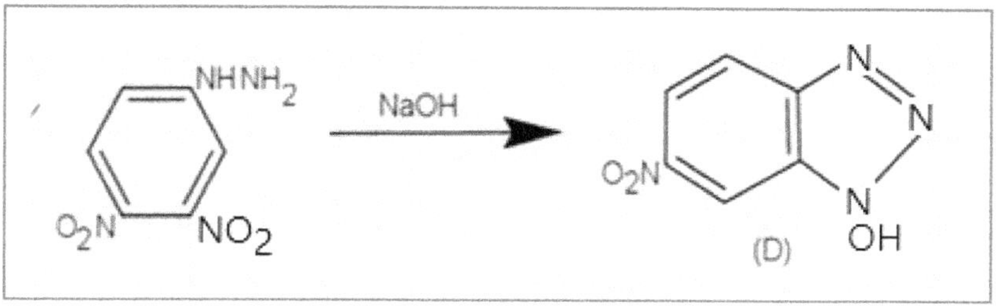

Benzotriazoles can also be obtained by **Ring Contraction** of bicyclic compounds, as 1,2,4–benzotriazin–3(2H)–one (E) in the presence of chloramine as aminating oxidant.

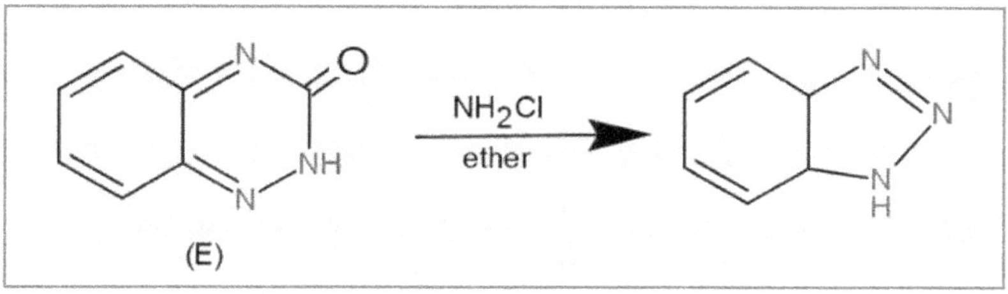

Various cyclization reactions lead to the formation of 1,2,4–triazoles in which oxadiazoles is many–a–times obtained as minor product.

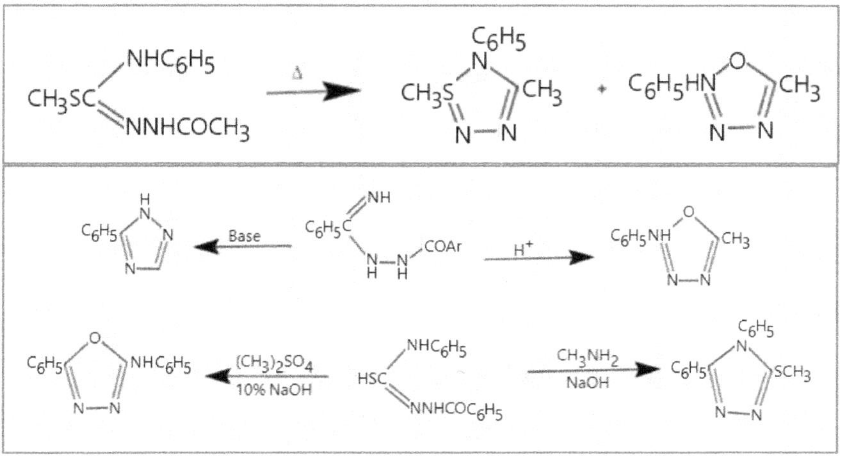

Oxadiazole is converted to triazole in the presence of ammonia and methylamine.

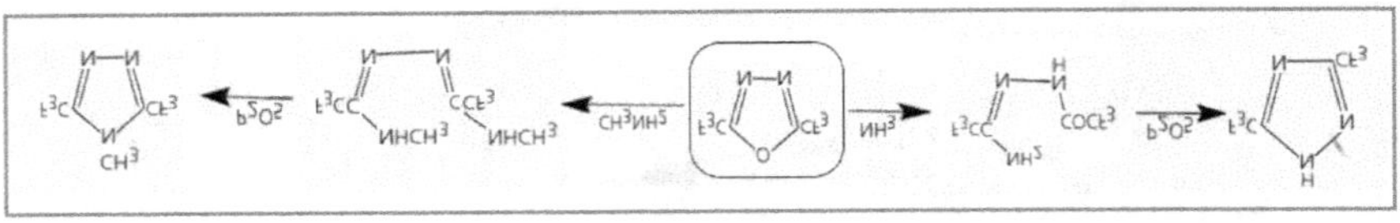

192

Triazoles with hydroxyl side chain can be prepared by hydrazinolysis of 1,3–oxazolin–4–one (F).

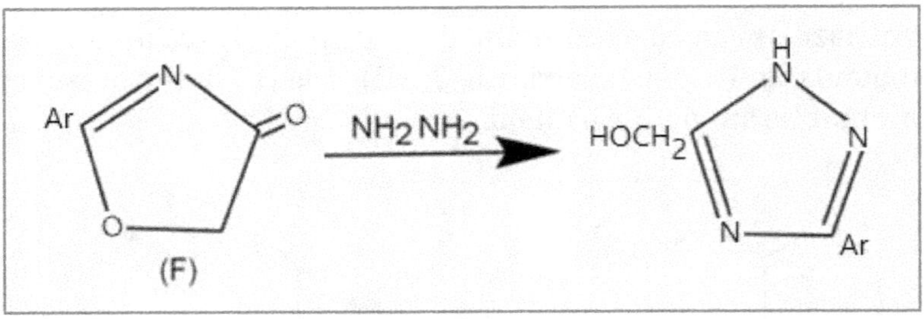

$$NH_2NH_2$$

(F)

Symmetrical 3,5–disubstituted 1,2,4–triazoles (G) can be prepared by self–condensation of nitrilimine (intermediate) in the presence of nitrile reagent.

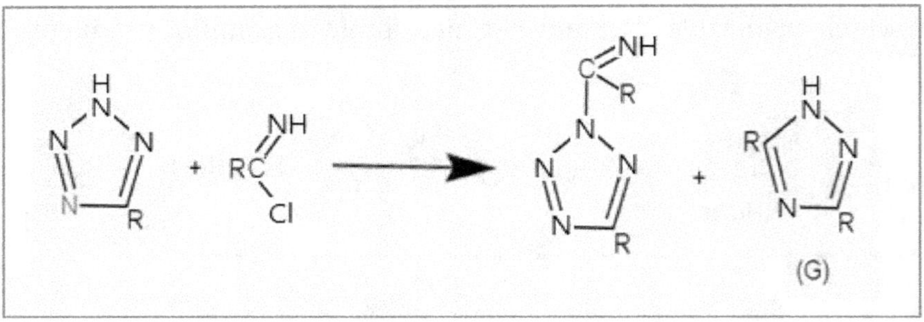

(G)

## PROPERTIES

N–methylation of 4–phenyl-1H-1,2,3–triazole takes place in the presence of dimethyl sulphate to produce 1–methyl and 2–methyl products.

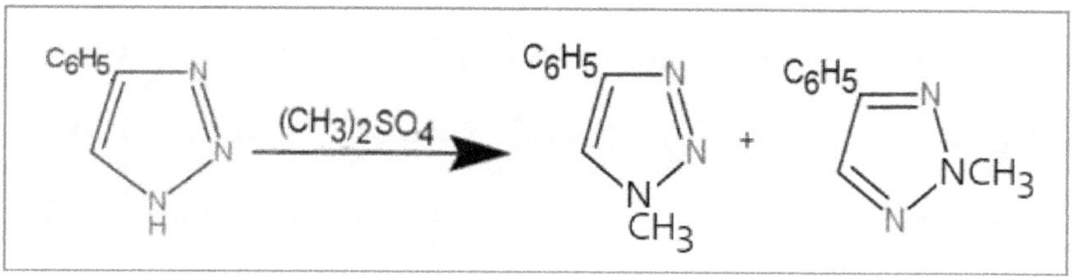

Benotriazoles show **acylation** on 1 – (or 3-) nitrogen atom in the presence of phenyl isocyanate.

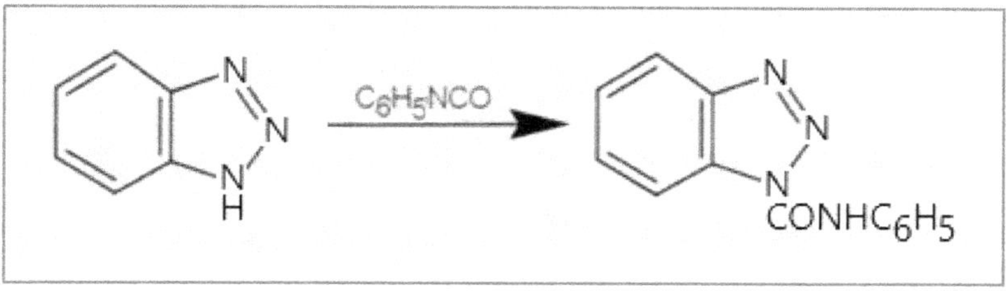

In the presence of potassium permanganate, by **direct oxidation**, aromatization is brought about.

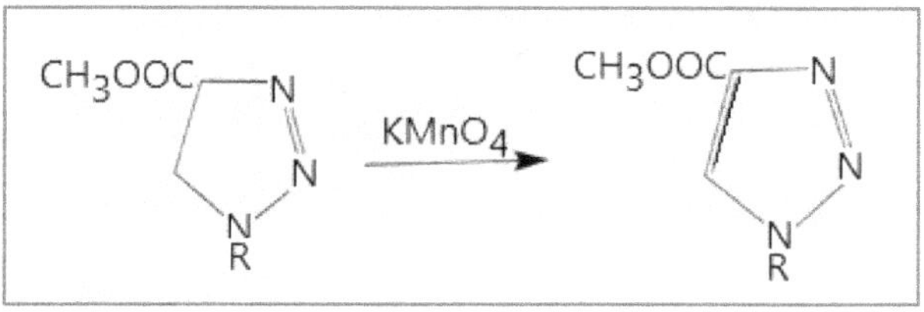

Triazolinones are easier to clear in the presence of acids.

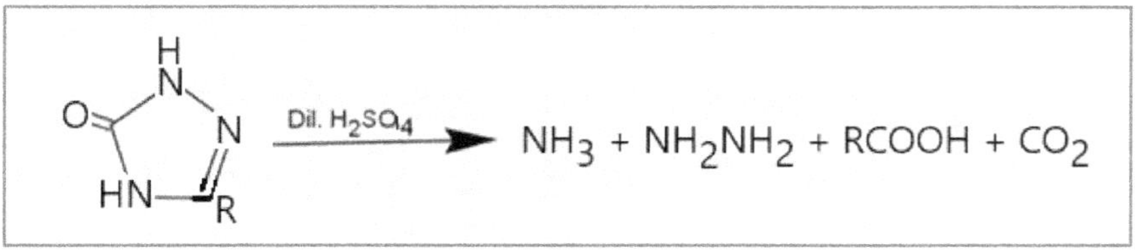

**Quaternization** of 1,3,5 – trimethyl – 1,2,4 – triazole occurs in the presence of methyl iodide.

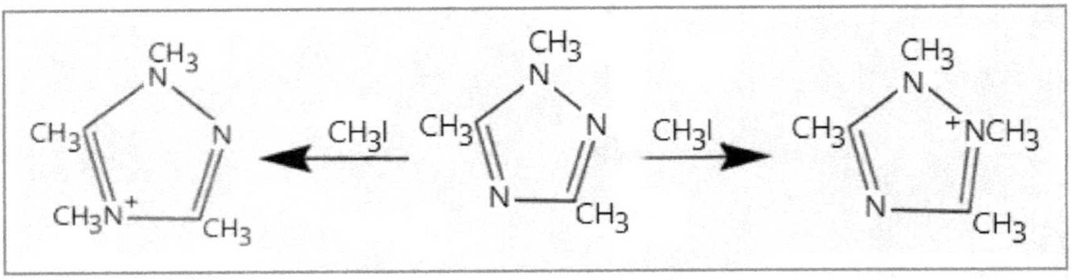

194

Triszolinediones show N – alkylation involving addition of cyclic or acyclic -C=CH to the former.

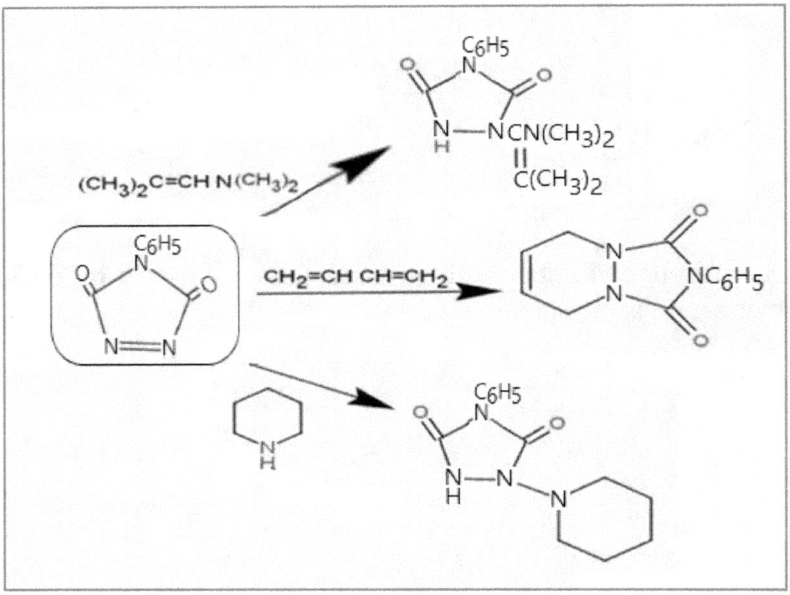

1,2,4 – triazolidine – 3,5 – dione gets diacetylated easily.

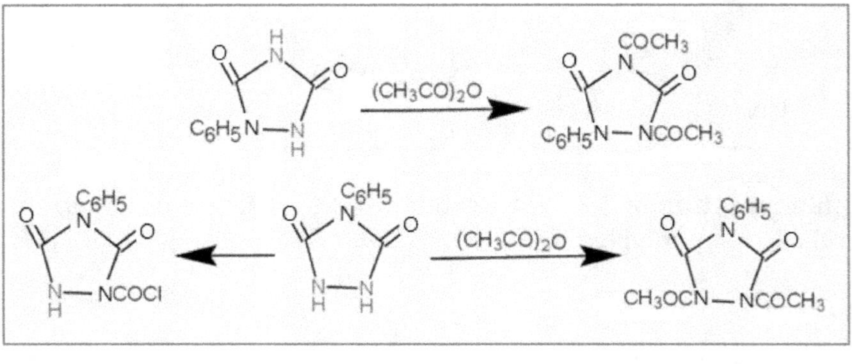

## APPLICATIONS

- 2–Phenyl–1,2,3–triazole–4–carboxylic acid **inhibits growth** in *Staphylococcus aureus, Escherichia coli and Salmonella typhi.*

- Chloro–substituted benzotriazoles (A) & (B) show **anti-inflammatory** activities.

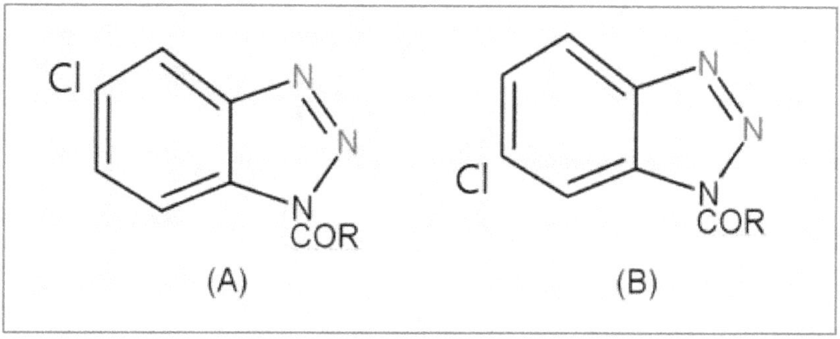

(A)

(B)

- **Bactericidal** activities are shown by IH– and 2H–1,2,3–triazole nucleosides like (C) & (D).

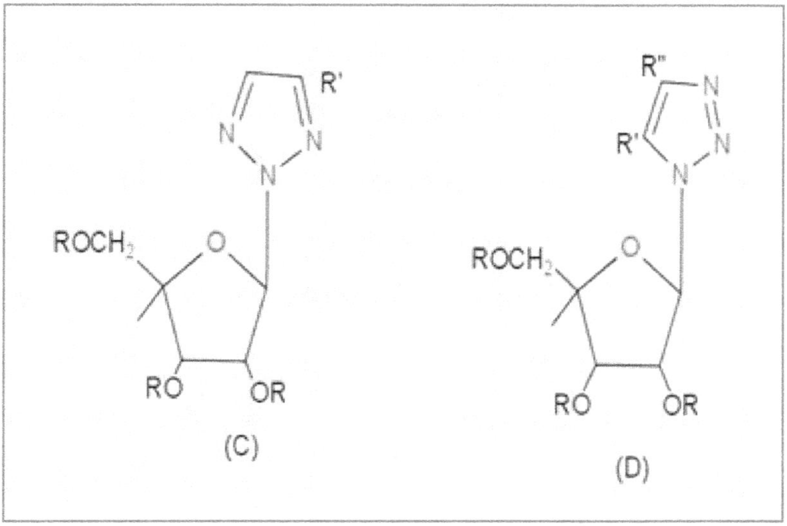

(C)

(D)

- Chloroform solution of 1,2,3–benzotriazole is used for the extraction of iron (II), iron (III), copper (II) & manganese (II).

- 1,2,3–Triazole and benzotriazoles have been used as **photo-stabilizers** for fibres, plastics or dyestuffs, for e.g. – (E) & 2–phenyl–2H–1,2,3–triazoles (F).



- 5–Methyl and 5-ethyl derivatives of 1-(3-tri fluoromethyl)-5–phenyl-1,2,3- triazole show strong non-selective **herbicidal** activity.



- There are certain compounds with core structures of 1,2,3-triazoles (a) which **regulate** local plant growth.

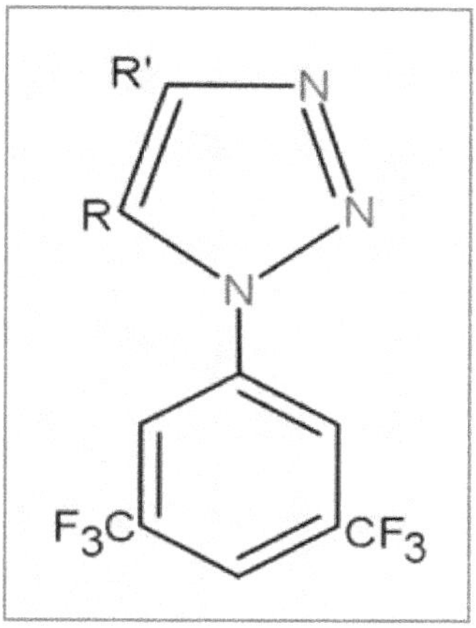

- **Anti-inflammatory** properties have been reported in simple thione.

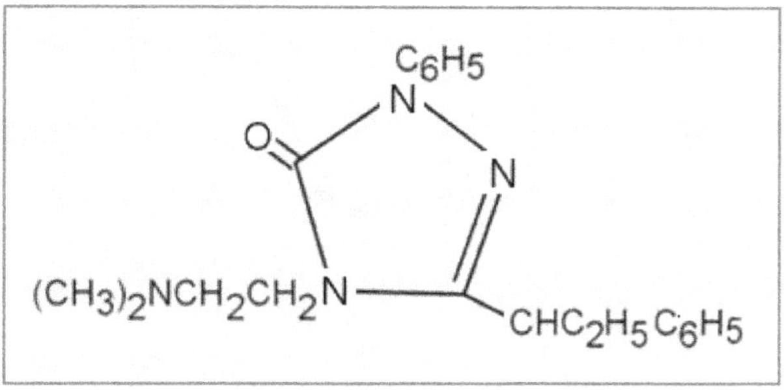

## 7.6.3 TETRAZOLES

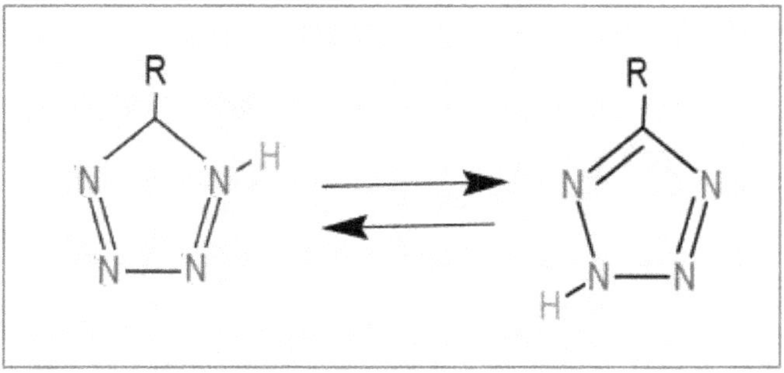

198

## SYNTHESIS

Tetrazole derivations can be obtained by the reaction of 3–diazoamino–5-methyl isoxazole (A) with base.



With when 5–arylamino thiatriazole(B) is heated base, it gives tetrazoline–5-thiones.



2,5-Disubstituted tetrazoles are obtained by **rearrangement** of 1,2,3 triazole which have aryl- or unsaturated nitrogen systems.



5-Aminotetrazole is an important tetrazole derivatives and is used in several ways.

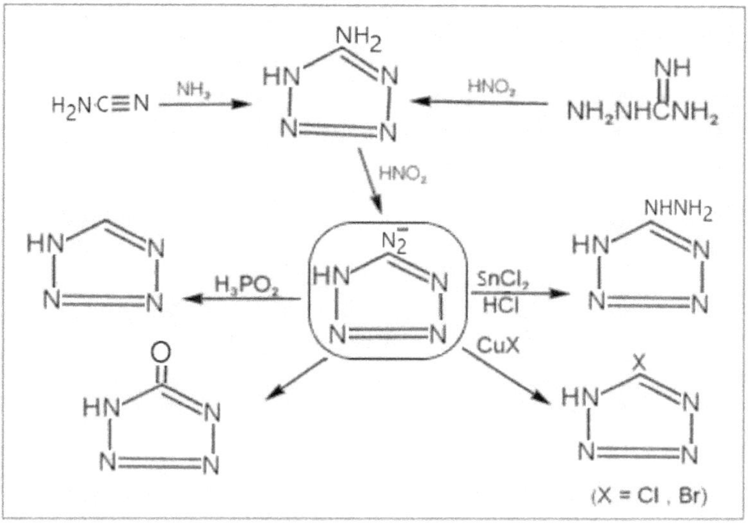

## PROPERTIES

**Photolysis** of 1,4-disubstituted tetrazole derivatives result in the elimination of nitrogen.

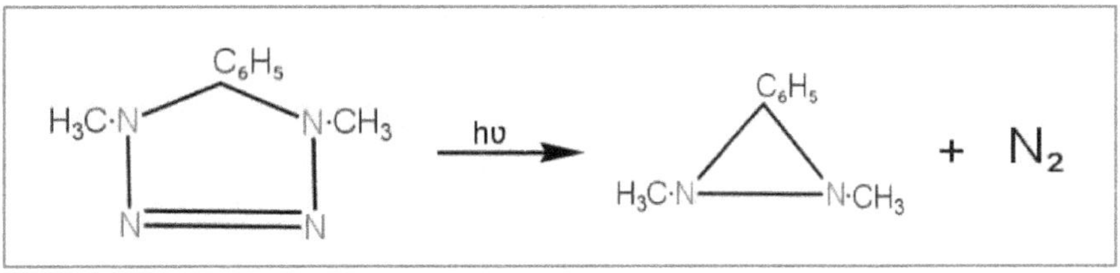

5-Phenyl-1-(2,6-dimethoxy phenyl) tetrazole(A) **disintegrates** into a variety of products either in the presence of light or heat.

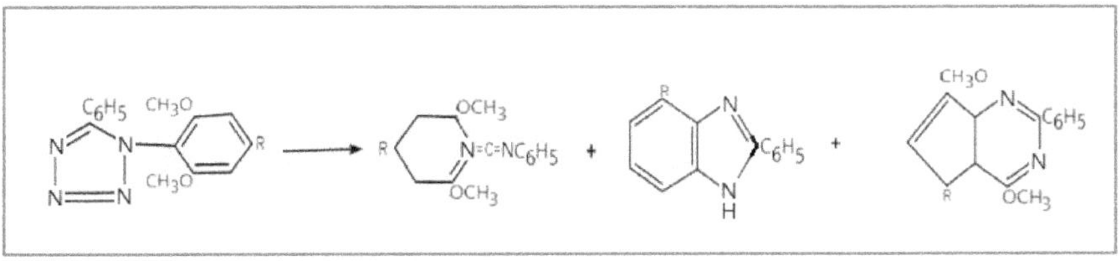

5-Aminotetrazole gave a variety of reactions with different reagents, with isocyanate it gave 1-(N-arylcarboxamido) tetrazole (B) while

200

with chloro-sulphonyl isocyanate it yielded 5-azido-2H-1,2,4,6-thia-triazin-3(4H)-one(C).

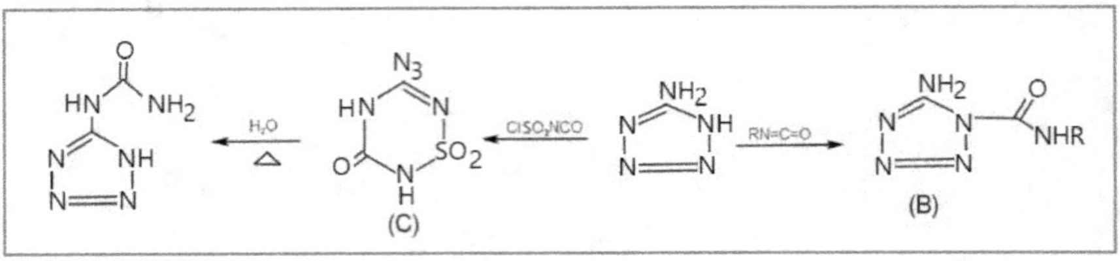

5- Aryloxy-1-Phenyl tetrazole on **reductive cleavage** results in the removal of only group.

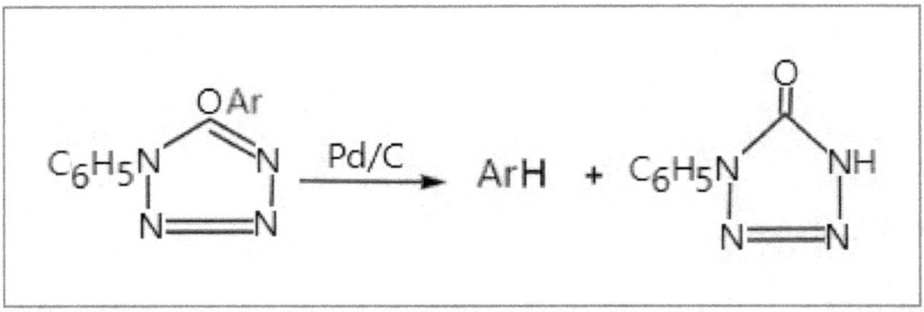

1- Aminotetrazole gives various reactions with sulphonyl halides, aldehydes, isocyanate etc.to yield a variety of products.

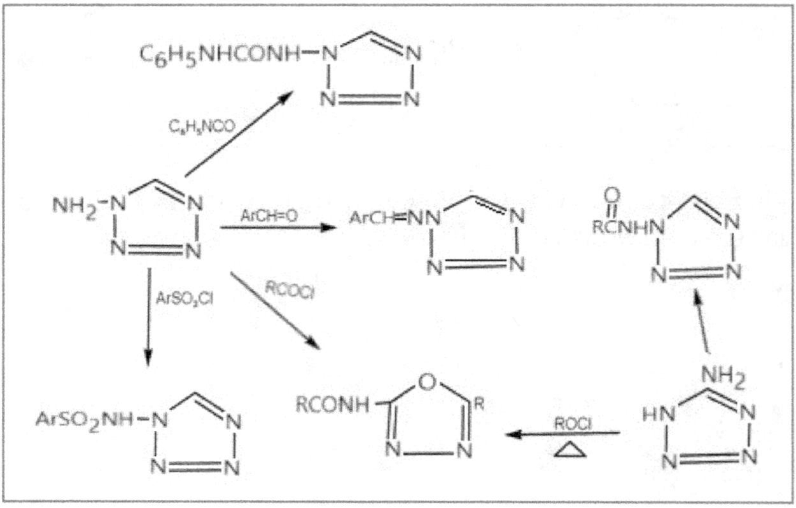

## APPLICATIONS

1. **Anti - inflammatory** property is shown by tetrazole alkane acid-
(A) **5-** phenyltetrazole(B).

(A)     (B)

2. Carboxy group at position-2 in chromone system shows anti-aller-
genic property and is found useful in bronchial asthma. Same type of
activity is found in chromone having N-unsubstituted tetrazole group
at C-2, C-6, C-8.

## REFERENCES

Boyd, G.V. and Wright, P.H. (1972); J. Chem. Soc; Perkin Trans., 1, 909.
Boyd, G.V. and Wright, P.H. (1972); J. Chem. Soc, Perkin Trans, 1, 914.
Schneiders, P., Heinze, J. and Baumgartel, H. (1973); Chem. Ber, 106, 2415.
Taylor, E. M. and Portnoy, R.C. (1973); J. org. Chem, 38, 806.

Lazaro, R., Mathiew, D., Fhan, R., Tan, L. and Elguero, J. (1977); Bull. Soc. Chim. Fr., 1163.

Van Vyre, T. and Viehe, H.G. (1974); Angew. Chem. 2nd Ed Engl., 1379.

Beam, C. F., Sandifer, R. F., Foote, R. S. and Hauser, C. R. (1976); Synth. Commun., 6, 5.

Begland, R. W. and Hartter, D. R. (1972); J. Org. Chem., 37, 4136.

Vieha, H. G. and Janousek, Z. (1973); Angew. Chem Int. Ed. Engl, 12, 806.

Potts, K.T. and Hussain, S. (1971); J. Org. Chem., 36, 10.

Taylor, E. C., Berchtold, G. A., Geockner, N. A. and Stroehmann, F. G. (1961); J. Org. Chem, 26, 2715.

Lozinskii, M. O., Shivanyuk, A. F. and Pelkis, P. S. (1971); Khim. Geterotsikl. Seodin, 471 (Chem Abstr 1972, 76, 25, 184)

Potts, K. T. and Armbruster, R. (1970); J. Org. Chem, 35, 1965.

Potts, K. T. and Armbruster, R. (1971); J. Org. Chem, 36, 1046.

Potts, K. T. and Kane, J. (1973); J. Org. Chem, 38, 3087.

Potts, K. T. and Kane, J. (1975); J. Org. Chem, 40, 2600.

Atkins, R. R. Glue, S. E. J. and kay, I. T. (1973); J. chem. soc. poleim tran, 1, 2644.

Pulst, M. and Weissentels, M. (1976); Z. Chem, 16, 337.

Takahash, M., Tan, H., Fukushimi, K. and Yamaohi, H. (1977); Bull. Chem. Soc. Jpn., 50, 2689.

Landquist, J. K. (1970); J. Chem. Soc. (C), 63.

Hauptmann, S., Wilde, H. and Moser, K. (1976); Tetrahedron Lett., 3295.

Hauptmann, S., Wilde, H. and Moser, K. (1971); J. Prakt. Chem., 313, 882.

Sieler, J., Wilde, H. and Hauptmann, S. (1971); Z. Chem., 11, 179.

Anderson, R. C. and Hsiao, Y. Y. (1975); J. Hetercycl. Chem., 12, 883.

Gagnoli, N. and Ricci, A. (1956); Ann. Chim. (Rome), 46, 275.

Grimmett, M. R. (1970); Adv. Heterocyl. Chem., 12, 103.

Schmidt, R. R. (1965); Chem. Ber., 98, 334.

Martin, D. and Weise, A. (1967); Chem. Ber., 100, 3736.

Schmidt, R. R. and Schwille, D. (1969); Chem. Ber., 12, 269.

Schmidt, R.R. (1965); Chem. Ber., 98, 334.

Hunig, S. and Hubner, K. (1962); Chem. Ber., 95, 937.

Yoneda, F. and Nagametsu, T. (1975); Bull. Chem. Soc. Jpn., 48, 1484.

Tomiwatu, Y. Satoh, K. and Sakamoto, M. (1977); Heterocycles, 8, 109.

Reid, S. T. (1970); Adv. Heterocycl. Chem., 11, 1.

Wentrup, C. (1981); Adv. Heterocycl. Chem., 28, 231.

Grimmett, M. R. (1980); Adv. Heterocycl. Chem., 27, 241.

Micetich, R. G. (1970); Can. J. Chem., 44, 2006.

Clapp, L. B. (1976); Adv. Heterocycl. Chem., 20, 65.

Timpe, H. J. (1974); Adv. Heterocycl. Chem., 17, 218.

Wakefield, B. J. and Wright, D. J. (1979); Adv. Heterocycl. Chem., 25, 147.

Wunsch, K. H. and Boulton, A. J. (1967); Adv. Heterocycl. Chem., 8, 277.

Lakhan, R. and Ternai, B. (1974); Adv. Heterocycl. Chem., 17, 99.

Weinstock, L. M. and Pollak, P. I. (1968); Adv. Heterocycl. Chem., 9, 107.

Burger, K., Schickaneder, H. and Meffert, A. (1975); Z. Naturforsch.; Teil B, 30, 622.

Migliara, O. and Sprio, V. (1981); J. Heterocycl. Chem., 18, 271.

Crenshaw, R. R., Luke, G. M. and Siminoff, P. (1976); J. Med. Chem., 19, 262.

Grimshaw, J. and Desilva, A. P. (1980); Can. J. Chem., 58, 1880.

Denzel, T. (1974); Arch. Pharm., 307, 177.

Senga, K., Novinson, T., Wilson, M. R. and Robins, R. K. (1981); J. Med. Chem., 24, 610.

Reisch, J. and Ossenkop, W. F. (1973); Chem. Ber., 106, 2070.

Reisch, J. and Fitzek, A. (1974); Arch. Pharm., 307, 211.

Perez, J. D. and Yranzo, G. I. (1987); J. Org. Chem., 47, 2221.

Schofield, K., Grimmett, M. R. and Keene, B. T. R. (1976); Heteroaromatic Nitrogen Compounds: The Azoles', Cambridge University Press, Cambridge.

Wiley, R. H. and Wiley, P. (1964); Chem. Heterocycl. Compd., 20, 1.

Mityurina, K. V., Kharchenko, V. G. and Cherkesova, L. V. (1980); Chem. Heterocycl. Compd., 16, 180.

Auzzi, G., Cecchi, L., Costanzo, A., Vettori, L. P. and Bruni, F. (1978); Farmaco. Ed. Sci., 33, 14.

Hauser, M., Peters, E. and Tieckelmann, H. (1961); J. Org. Chem., 26, 451.

Ramalingam, K., Thyvelikakath, G. X., Berlim, K. D., Chesnut, R. W., Brown, R. A., Durham, N. N., Ealick, S. E. and Van der Helm, D. (1977); J. Med. Chem., 20, 847.

The Merck index (1976); ed.M. windholz, mercle and co. Rahway. New jersr, 9th Edn.

Burger's Medicinal Chemistry (1980); Wiley, New York, 4th edn.

Elguero, J. and Marzin, C. (1926); Adv. Org. Chem., 9(**1**), 533.

Engel, N. and Steglich, W. (1978); Liebigs Ann. Chem., 1916.

Grimmett, M. R. (1980); Adv. Heterocycl. Chem., 27, 241.

Pozharskii, A. F., Garnovskii, A. D. and Simonov, A. M. (1966); Russ. chem. Rev. (Engl. Transl.), 35, 122.

Popov, I. I., Bogachev, Yu. G., Tkachenko, P. V., Simonov, A. M. and Tertov, B. A. (1976); Chem. Heterocycl. Compd. (Engl.transl.), 12, 437.

Nakano, T., Rodriguez, W. de Roche, S. Z., Larrauri, J. M., Rivas, C. and Perez, C. (1980); J. Heterocycl. Chem., 17, 17.

Usaevich, Yu. Ya., Fel'dman, I. kh. and Boksiiner, E. P. (1971); Chem. Heterocycl.

Compd. (Engl. Transl.), 7, 749.

Kasina, S. and Nematollahi, J. (1975); Synthesis, 162.

Popov, I. I. and Simionov, A. M. (1974); Chem. Heterocycl. Compd. (Engl . transl.), 10, 1491.

Popov, I. I. and Simionov, A. M. and Zubenko, A. A. (1975); Chem. Heterocycl. Compd. (Engl .transl.), 11, 123.

Schatnev, P. V., Shvartsberg, M. S. and Bernshtein, I. Ya. (1975); Chem. Heterocycl. Compd. (Engl. Transl.), 1975, 11, 718.

Pozarskii, A. F., Zvezdina, E. A., Kashparov, I. S., Andreichikov, Yu. P., Maryanovskii, V. M. and Simonor, A. M. (1971); Chem. Heterocycl. Compd. (Engl. Transl.), 7, 1156.

Nagai, T., Fukushima, Y., Kuroda, Shimizu, T. H., Sekiguchi and Matsui, K. (1973); Bull. Chem. Soc. Jpn., 46, 2600.

Nagarajan, K., Nair, M. D. and Desai, J. A. (1979); Tetrahedron Lett., 53.

Yoneda, F., Higuchi, M., Matsumara, T. and Senga, K. (1973); Bull. Chem. Soc. Jpn., 46, 1836.

Yamazaki, A., Kumashiro, P., Takenishi, T. and Ikehara, M. (1968); Chem. Pharm, Bull, 16, 2172.

Cook, A. H. and Thomas, G. H. (1950); J. Chem. Soc., 1884.

Falco, E. A., Elion, G. B., Burgi, E. and Hitchings, G. H. (1952); J. Am. Chem. Soc., 74, 4897.

Chin, A., Hung, M. H. and Stock, L. M. (1981); J. Org. Chem, 46, 2203.

Fujii, T., Sato, T. and Itaya, T. (1972); Chem. Pharm. Bull., 49, 1731.

Baines, K. M., Rourke, T. W. and Vaughan, K. (1981); J. Org. Chem., 46, 856.

Bleiholder, R. F. and Shechter, H. (1968); J. Am. Chem. Soc., 90, 2131.

Wrubel, J. and Mayer, R. (1979); Z. Chem., 19, 446.

Rees, C. W. and Sale, A. A. (1971); Chem. Commun, 1971, 532.

Herbst, R. M. and Klingbeil, J. E. (1958); J. Org. Chem., 23, 1912.

Hergenrother, P. M. (1969); J. Heterocycl. Chem., 6, 965.

Simiti, I. And Ghiran, D., (1971); Farmacia (Bucharest), 19, 199 (Chem. Abstr, 1971, 75, 76, 692).

Temple, Jr, C. (1981); Chem. Heterocycl. Compd. ,87, 72.

Leistner, S., Wagner, G. and Richter, H. (1974); Z. Chem., 14, 267.

Brown, H. C. and Kasal, R. J. (1967); J. Org. Chem., 32, 1872.

Raap, R. (1972); Can. J. Chem., 49, 1792.

Kadaba, P. K. (1978); Synthesis, 694.

Temple Jr., C. (1981); Chem. Heterocycl. Compd., 37, 530.

Nadkarny, V. V., Rao, R. S. and Fernandes, P. S. (1976); J. Indian Chem. Soc., 53, 833.

Kreutzberger, A. and Stratmann, J. (1980); J. Heterocycl. Chem., 17, 1505.

ICN Pharmaceuticals Inc., (1976) U. S. Pat. 3, 968, 103 (Chem.Abstr. 1976, 85, 94, 660).

Simonenko, V. I., Pyatnitskii, I. V. and Nizkaya, G. K. (1980); Ukr. Khim. Zh. (Russ. E. D.), 46, 1099. (Chem.Abstr, 1981, 94, 53, 670).

Guillet, I. E. (1972); Pure Appl. Chem., 7230, 135.

GAF Corp. (1972) Ger. Pat. 2., 864, (Chem. Abstr. 1972, 76, 99,674).

Massori, V., Baldi, L. and Biarchetti, G. (1977); Chim. Ind. (Milan), 59, 438.

Sandoz Ag, (1981) Braz. Pat. 81 01 239 (Chem. Abstr., 1982, 96,69, 006).

Butter, R. N. (1977); Adv. Heterocycl. Chem., 21, 322.

Benson, F. R. (1967); Heterocyclic Compounds; Ed. Elderfield; Wiley, New York, Vol. 8, 1.

Akiyama, T., Kitamura, T., Isida, T. and Kawanisi, M. (1974); Chem. Lett., 185.

Gilcwrist, T. L., Moody, C. J. and Rees, C. W. (1976); J. Chem. Soc., Chem. Commun., 414.

Denny, G. H., Cragoe, Jr, E. J., Rooney, C. S., Springer, Hirshfield, J. M. and Mc Cauley, J. A. (1980); J. Org. Chem., 45, 1662.